REAL GREEN

Rethinking Political and International Theory

Series Editors:
Keith Breen, Dan Bulley and Susan McManus,
all at Queens University Belfast, UK

Committed to a critical and creative exploration of the ways that canonical approaches in political and international theory may be applied to 21st century politics, this series presents pioneering theoretical work on contemporary political issues that both furthers our understanding and shapes exciting new agendas for research. The works featured will advance our appreciation of the relevance of seminal thinkers to the current socio-political context, as well as problematize, and offer new insights into, key political concepts and phenomena within the arena of politics and international relations.

Also in the series

The Politics of Misrecognition
Edited by Simon Thompson and Majid Yar
ISBN 978 1 4094 0169 8

Institutionalizing Agonistic Democracy
Post-Foundationalism and Political Liberalism
Ed Wingenbach
ISBN 978 1 4094 0353 1

Power, Judgment and Political Evil
In Conversation with Hannah Arendt
Edited by Andrew Schaap, Danielle Celermajer and Vrasidas Karalis
ISBN 978 1 4094 0350 0

Real Green
Sustainability after the End of Nature

MANUEL ARIAS-MALDONADO
University of Málaga, Spain

Routledge
Taylor & Francis Group

LONDON AND NEW YORK

First published 2012 by Ashgate Publishing

2 Park Square, Milton Park, Abingdon, Oxon OX14 4RN
711 Third Avenue, New York, NY 10017, USA

Routledge is an imprint of the Taylor & Francis Group, an informa business

First issued in paperback 2016

British Library Cataloguing in Publication Data
Arias-Maldonado, Manuel.
  Real green : sustainability after the end of nature. --
  (Rethinking political and international theory)
  1. Human ecology. 2. Political ecology. 3. Green movement.
  I. Title II. Series
  320.5'8-dc22

Library of Congress Cataloging-in-Publication Data
Arias-Maldonado, Manuel.
  Real green : sustainability after the end of nature / by Manuel Arias-Maldonado.
      p. cm. -- (Rethinking political and international theory)
  Includes bibliographical references and index.
  ISBN 978-1-4094-2409-3 (hardback)
  1. Political ecology. 2. Green movement. 3. Sustainable development. I. Title.
  JA75.8.A74 2012
  320.5'8--dc23

                                                                  2011032304

ISBN 978-1-4094-2409-3 (hbk)
ISBN 978-1-138-24955-4 (pbk)

To my mother and sisters

Contents

Our grandfathers were less-housed, well-fed, well-clothed than we are. (...) The gadgets of industry bring us more comforts than the pigeons did, but do they add as much to the glory of the spring?

Aldo Leopold

Mankind's old greatness was created in scarcity. But what may we expect from plenitude?

Saul Bellow

Ramsey was a bourgeois thinker. That is, his thought was meant to order a given community.

Ludwig Wittgenstein

Foreword

A book can be said to be, ultimately, a mere *opinion*. Of course, it is a verifiable opinion when it belongs to the natural sciences and a grounded opinion when it belongs to the social sciences – but an opinion it remains. The author has reflected upon a given subject and formed a particular view of it, which he expresses as best as he can and in as many pages as the patience of the reader can endure. If there were, so to speak, a natural conclusion for every subject, seven books would suffice. But there is not. Yet, on the other hand, there are reasons to object that the book belongs to the author who signs it. As every book is the result of a sum of disparate influences, it is also, unavoidably, a simplification. Years of work are condensed into 200 pages and these pages never seem to say what we wanted them to say. It is all part of the job. However, these little tragedies are immediately forgotten once the author contemplates the first printed copy: the book seems truly *ours*. The only thing to do then is to think of its genealogy, modest history and unwilling appropriations. Signing it is bold enough!

The text I dare to present here has been written during the last year, but it is the outcome of a whole decade of interest and research. I do not claim to have read *everything* that is to be read upon this wide topic, but I hope that I have read *enough* for the book to be a contribution to the ongoing debate about sustainability and environmentalism – that is, about the shape of the coming society. My main contention is that environmentalism should be reframed along more realistic lines, abandoning the language of crisis and embracing instead a more pragmatic, liberal, and pluralistic view of sustainability. The main reason for doing so is the fact that nature does not exist anymore, because the human dominion of nature has deprived it of its independence and turned it into human environment. That is not to say that the remaining natural world should not be protected – it should. But a thorough reorientation of green politics is needed if we wish it to be influential in the years to come. Needless to say, it is for the reader to decide whether I have succeeded or failed in defending my case.

I would like to thank a number of people who have generously contributed to this book in different ways. My Spanish colleagues have been a constant example throughout the years and I wish I could mention all of them. Angel Valencia introduced me to this fascinating field of study and has remained an inspiration ever since. Fernando Vallespín and Rafael del Águila, who unfortunately is no longer among us, have kindly supported me with guidance in the academic realm whenever I have needed it. Sebastián Escámez and José Manuel Cabra are daily interlocutors in matters of life and science and have helped me to keep my spirits high at all times.

Elena García-Guitián, Joaquín Abellán, Francisco Llera, Ramón Máiz and Carlos Alba have also been very helpful, for various reasons, in recent years.

On the supranational front, I would specially like to thank thank Andrew Dobson for his advice and support. I began studying these issues in Keele University some time ago and without his hospitality I would have not survived that British winter! John Barry, Marcel Wissenburg, Carme Melo, Derek Bell, Brian Doherty, Sherilyn McGregor, Martin Baumeister, and Volkmar Lauber have also been generous to me on different occasions during the period in which this book was being researched and written. And the same goes for Michael Watts and Steven Weber, former directors of the Institute for International Studies at the University of Berkeley. However, this book has actually been finished during a stay at the Rachel Carson Center in Munich, whose directors, Christoph Mauch and Helmut Trischler, have treated me with a splendid generosity for which I am very thankful. This gratitude is to be extended, in fact, to all my colleagues in this excellent institution.

On the other hand, I also wish to acknowledge the excellent work done by my two successive editors at Ashgate, Natalja Mortensen and Margaret Younger. In this same regard, Professor John M. Meyer has produced a critical reading of the manuscript which has contributed to improving it – I really appreciate it. I would also like to thank Alicia Shelley for correcting and polishing my written English. However, I have translated quotations from German and Spanish into English. Needless to say, the mistakes and flaws the book may possess are my sole responsibility.

On a more personal note, I would like to acknowledge and thank the support that I have received during the completion of this book from my family and friends. My father died prematurely three years ago and his absence is still – and will ever be – deeply felt. I wish he could have lived to enjoy this publication. My mother and my two sisters, Elisa and Begoña, have had patience with my impatience and I give them all the credit for it. As for my friends, from Jose Luis and Fran and Victoria to Jaime and Cristian, they are a family of choice without which life would simply be boring.

                                                  Manuel Arias-Maldonado

Introduction:
An Imaginary Crisis?
Reframing Environmentalism

The Ecological Crisis and its Metaphors

There is a well-known story about an outsider who tries to find his way in a rural town where he has got lost. He asks one of the locals for directions, only to hear the following answer from an old man: "If I wanted to go there, I would not start from here!" And he says no more. Be it a true story or not, it resembles the way in which human society is facing the quest for environmental sustainability: we are not sure whether we truly wish to get there, nor if we have chosen the best starting point to do so. Or so it seems. Yet it could also be the case that both means and ends are confusing because we have not realised what sustainability is about, i.e. we do not know *where* are we heading to, nor *why* are we doing so. In other words, what we lack is an adequate understanding of the relationship between humanity and nature, one that prefers prose over poetry. So far we have chosen the wrong narrative – guilt instead of sympathy – about our performance as a species, wherefrom the wrong framing for sustainability – survival instead of refinement – has arisen. This should be challenged. And that is what this book is about: the philosophy and politics of a realistic sustainability. After all, we must get there anyway.

However, maybe it is a mistake to pose the question in terms of a future path leading at some point to a sustainable society. It could also be argued that sustainability is already happening. Since climate change arrived, we have all become greens. Of course, some are greener than others. But it is undeniable that the relationship between society and nature has never been so conspicuous in the public arena, nor has it ever been so present in the policy agenda, as it is now. We are witnessing the consolidation of nature as a political problem and as a social concern. From fashion miles to going vegetarian, passing through renewable energy and the protection of forests, all sort of green issues make their way into people's lives and attract the attention of the media. No matter how rhetorical or clumsy some of those attitudes may be, nobody seems inclined to say a word *against* sustainability anymore. But why should they? Taking care of the planet, saving humanity – these are edifying tasks. Look at those penguins!

Still, such cultural success does not entail that the public debate about the environment is taking place in the right terms. Many environmentalists would concur, since a greater stress on our moral duty to nature, as well as on the subsequent need to radically transform the current economic and social organisation, would be desirable for them. In sum, they see the current lexicon as too anthropocentric to make a difference. But what if it is the other way around? What if it is precisely the predominance of some old green tenets that should be questioned as a basis for the public conversation on sustainability? Let us consider the following statement:

> For, if current trends are allowed to persist, the breakdown of society and the irreversible disruption of the life support systems on this planet, possibly by the end of the century, certainly within the lifetime of our children, are inevitable.

A warning of this kind is very common nowadays. However, it belongs to a well-known pamphlet published in 1972, aptly entitled *Blueprint for Survival* (Goldsmith et al. 1974: vii). The fact that the prediction has not been fulfilled does not seem to be a reason not to insist on it. In fact, such obstinacy is arguably one of the reasons why this style of reasoning left the pamphlets and reached the official reports. As Al Gore, former vice-president turned green activist, told American Congress on March 2007: "Mankind faces a planetary emergency, a crisis that threatens the survival of our civilisation and the habitability of the Earth". It is the same tune, played by a different orchestra. And now the audience is humming it.

The rise of nature as a widespread concern has thus been achieved in the very terms posed by the green movement since its inception more than 40 years ago: the fear of ecological disaster. A simple premise, followed by a categorical consequence: humanity faces a global ecological crisis which demands a thorough transformation of our social organisation, in order to set up a new relationship with nature based on the moral recognition of its value. Besides, this change is to be made urgently: global warming puts a time-bomb on the table that make all previous green predictions pale in comparison. Henceforth, we are "a generation at a crossroads", obliged to make an unparalleled ethical turn: "The moral, the practical, the ultimate question is what we will do about it, what we will give up today so that we and our descendants and the rest of life on earth may have their tomorrow" (Jackson 2002: 132). A moral urgency translated into political urgency: the social order must adapt itself to nature as quickly as possible. Forty years ago, only a few visionaries saw it that way; today, it almost looks like a cliché.

Be that as it may, the ecological crisis is no less important than its metaphors. According to an old literary trick, a sick body can serve as a physiological expression of the patient's moral or emotional unrest. The same goes for the ecological crisis, which has been constantly depicted as a crisis of civilisation. The material side of socionatural relations is thus linked to a deep symbolic dimension, whose main corollary is that Western society has reached its limits. *Rien ne va plus*. Such a collapse is anticipated by contemporary imagination through novels, films, comic-books: a culture dreaming up its end. It is this meaning of the ecological crisis that pervades the whole environmental debate.

The labels are countless. It has been said that this is a "crisis of culture and character" (Eckersley 1992: 17), a "crisis of inaction" (Porritt 1984: 116), as well as a "crisis of perception" that is "social rather than natural" (Capra 1996: 13; Saurin 2001: 65). It has even been suggested that it is precisely this symbolic dimension that counts, since we face "first and foremost a crisis of consciousness" (Blühdorn 2000: 37), a "crisis of objectivity" (Latour 2004: 20), a "crisis or failing of reason and culture" (Plumwood 2002: 15), although we can also read that it is "first and foremost, a crisis of *governance*" (Adger and Jordan 2009: xvii). Indeed, a crisis of consciousness *should* lead to a crisis of governance; and vice versa. The advent of climate change has deepened this rationale, reinforcing the link between human failure and ecological collapse, with the former revealing a "long-term systemic crisis" (Derber 2010: 1). In fact, global warming has provided a new, encompassing theme for framing disaster. Since the German journal *Der Spiegel* used the expression *Klima-Katastrophe* for the first time in April 1986, the image of a planet gone climatically mad has become ubiquitous, replacing the nuclear threat as our most beloved apocalypse. The religious language of doom and judgement is mixed up with a language of acceleration and irreversibility, whose sharpest manifestation is the hypothesis of a "tipping point" climate catastrophe, i.e. a sudden collapse into a boiling Earth (Retallack and Lawrence 2007: 55; Pearce 2007). Apparently, there is no way out.

Or is there? Naturally, an argumentative construction of this sort points to the historic dominance of a set of beliefs that have provoked a gradual separation between humanity and nature; a breach leading to our current flirtation with extinction. Hence history is read again, this time as the incubation period of a spiritual illness leading to ecological crisis. Moreover, given the underlying attitudes, the crisis will allegedly have been reached *for certain* sooner or later (Black 1970: 18). Of course, it is this breadth of vision that makes environmentalism an encompassing ideology that goes far beyond ecology into moral and politics. It is not just to do with the environment, but also with the way in which we live and our socio-economic organisation, a point already stressed by early environmentalism (Ehrlich 1969; Meadows and Meadows 1972; Commoner 1971). At the same time, it is an astute formulation. If the ecological crisis is the final outcome of a sick civilisation, or even of a sick species, the cure cannot be superficial; it must go deep into the bone. An object devoured by its metaphors.

Such sense of urgency concerning survival is also meant to provide a strategic advantage, since we are not supposed to risk an eco-catastrophe. This line was also adopted by early environmentalism and has become a classic ever since: as market capitalism and liberal democracy are the culprits of the ecological crisis, as well as the main obstacles for dealing with it, we can only reach a sustainable state overcoming them. Subsequent proposals aimed at best to a utopian transformation of the social system, at worst to the imposition of an eco-authoritarian regime (see Hardin 1977a, 1977b; Heilbroner 1975; Ophuls 1977; Goldsmith et al. 1972). This latter option was more *consistent* than it seems: an ecologically enlightened technocracy can be seen as the only chance for survival, insofar as the radical measures we are in need of are both urgent and unpopular, i.e. they are to be imposed. Thus environmentalists were portrayed as the "vanguard for a new society" (Milbrath 1984). Needless to say, green theorists have long abjured from this stance, committing themselves to democratic values. Yet there is an insoluble tension between environmentalism and democracy, or at least between environmentalism and liberty, which actually *explains* that eco-authoritarianism happened to be a theoretical green horizon just once. In fact, this idea – catastrophism leading to exceptionalism – is being heard again in the call to arms against global warming.

However, for all the pedagogical virtues that exaggeration may possess, the environmental debate must not necessarily be founded upon the premise of the ecological crisis. There is an alternative. It could even be the case that the current distance between public *opinions* and individual *behaviours* in regard to sustainability has to do with the flaws of such rhetoric of doom. Of course, as a foundation for sustainability, most greens would prefer to underline human duties towards nature, instead of the threat to human survival. As it happens, fear is a more powerful device than the abstract call to moral extensionism, so from the outset a *public* emphasis on anthropocentric reasons has been considered a more profitable cultural and political strategy, notwithstanding the *private* emphasis on more ecocentric arguments (see Dobson 2000a: 21). It could well be the case that such arguments are the only ones available when trying to spread ecological values among the world population. However, environmentalists *do* believe in the ecological threat to survival, as their stance on global warming plainly shows. But none of that means that they are the *right* arguments.

On the contrary, framing our environmental problems as a pervading ecological crisis may have harmful implications. Every aspect of the environmental debate – from genetically modified food to biodiversity, passing through energy and conservation – are over-determined by the menacing ghost of the ecological crisis. As we shall see, this leads to a misguided understanding of socionatural relations, hence hindering the debate about a viable green society. No wonder that some sections of the green movement can be seen as defenders of anti-humanists or non-liberal positions – such as the need to suppress trade and travelling in order to live in one's own "bioregion", or the advisability of stopping immigration in the US for the sake of the environment (Sale 1985; Cafaro and Staples 2009). The main pathology of environmentalism has arguably been its refusal of the principle of reality, as expressed in the faith in a long gone nature. But this refusal doesn't have to be shared. In fact, environmentalism itself can and should change, in order to be a meaningful actor in the coming politics of sustainability. A commendable transformation that begins by questioning the tenet of the global ecological crisis.

Suddenly, Last Spring? Questioning the Ecological Crisis

The fact that the ecological crisis has become common place in contemporary culture does not necessarily validate its likeliness. On the contrary, this circumstance should warn us, given the ease with which received wisdom tends to spread through language, as a sort of cultural inertia that might or might not have real existence out of that culture. After all, if the ecological crisis can be read as a crisis of *consciousness*, it might also turn out to be an *imaginary* crisis. Surely an affirmation to be rigorously understood! It is not claimed that there are no environmental problems, nor that they do not deserve our careful attention – but it is indeed questioned that such problems constitute a *critical* situation, instead of a new episode in the intrinsically problematic relationship between society and nature.

In any case, an ecological crisis is not a category which can be directly deduced from the outer world. On the contrary, it is politically constructed upon certain philosophical foundations that project some *values* onto the observed *facts*. To begin with, it entails a value judgement on a series of *changes* that take place in the natural realm, or more exactly in the relationship between the former and society. This judgement is in turn dependent on a previous consensus about the desirability of a given, static *moment* in those relationships. Yet according to which criteria and appealing to which ultimate foundations? Are they religious, metaphysical, or secular? Now, this is a difficult question to be answered in a post-metaphysical age. If we respect moral pluralism regarding conceptions of nature and socionatural relations, the notion that we suffer an ecological crisis ceases to be self-evident. As Mike Hulme has stressed in connection to the discourses of fear surrounding climate change, "these discourses of fear are always situated – geographically, historically and culturally. They are not imposed by Nature, they are created through Culture" (Hulme 2009: 68). One person's worry is another person's indifference.

It could be suggested that *surviving* as a species can serve as a universal criterion for evaluating the state of the environment. If only because only religious conceptions of the good life can be realised in the hereafter! An ecological crisis could then be declared whenever the biophysical foundation for human life is seriously threatened. However, although it makes sense that some refined version of the survivalist argument is employed as a foundation for sustainability, it is not clear that a criteria derived wherefrom can also support the idea that we face a critical, unprecedented situation in socionatural relations. Indeed, some argue that the state of the environment is better than it was, if the term of comparison is the real world and not some idealised picture of a pristine nature or a better past. They belong to a tradition of thought that has run parallel to the evolution of environmentalism, claiming that human adaptability and ingenuity, together with the rise of a new ecological consciousness, have been increasing both human welfare and environmental quality, as a non-biased analysis of statistical data would come to show (see Maddox 1972; Simon and Kahn 1984; Lewis 1992;

Lomborg 2001; Simon 1981, 1995; Anderson 2004; Ridley 2010). Or maybe: an analysis as interested in an overall evaluation of the socionatural relations as in the state of the non-human world living within it. Needless to say, those defending the contrary would claim that the bias works precisely the other way around.

There is much to be discussed here. But if we concentrate first on the survival question, a general pattern is perceptible which brings two different views of the socionatural relations face to face: one is static, the other is dynamic. The classical environmentalist claims that nature will be exhausted by humanity if the current rhythm of depletion is maintained, since nature *is* exhaustible; the counterargument says that such relationship is intrinsically changeable, thanks to the human ability to reinvent it. Neither of these arguments can be proven, since they are bets on the future. However, the past tends to support the dynamic rather than the static view, as the bet between Julian Simon and Paul Ehrlich about the likely scarcity of natural resources illustrates. The story is well-known: Simon had predicted that natural resources were basically inexhaustible, whereas Ehrlich believed they were running out, so in 1980 they bet on the price of five metals selected by Ehrlich, who would win if their price had risen 10 years later, suggesting scarcity. The metals were copper, chrome, nickel, tin, and tungsten; by 1990, their price had fallen and Ehrlich lost the bet. Not that such failure can impress environmentalists, since they can produce arguments contesting the value of these episodes and claiming that absolute scarcity does not cease to be a possibility (see Lawn 2010a). However, this is not new: Malthus did not bet with Marx, because they were hardly coetaneous, but had he done so, he would have lost too.

It seems absurd to discuss that *so far* humanity has been able to find new ways of re-organising the socionatural relations in a durable – let us not say sustainable – way, despite the alleged occasional failure of *some* societies to do so (see Diamond 2006; Ponting 2007; McAnany and Yoffee 2010). Environmentalism has normally answered back that it is nonsensical to claim that, just because we have been able to solve such problems in the past, we are bound to do it again *now*. What has been termed the *cornucopian* or *prometean* tradition (Dryzek 2005: 51) would then be but a mere ideological reflection of a capitalist system that provoked the ecological crisis in the first place. In fact, it has been suggested, if optimism can be refuted by a global catastrophe, is it not the former, rather than the latter, that has to be carefully proven (Cotgrove 1982)? This rationale has been reinforced by global warming, a complex phenomenon whose precise evolution cannot possibly be predicted.

Nevertheless, betting on a current failure of humanity is less reasonable than expecting a confirmation of the historic pattern of success in the face of scarcity and overpopulation. It is interesting to note that environmentalism has traditionally defended a dynamic view of disaster together with a static view of human abilities, whereas the opposite is more probably true. The exponential development of current technologies makes the future state of human abilities thoroughly unpredictable, turning some green worries about that very *deep future* into delirious affirmations, e.g. the plea for a "quality survival" in 100,000 years' time from now (see Cocks 2003). There is probably an inescapable anthropological factor in play, namely, the human propensity to see their historic situation as unique and critical each time. In one of his early movies, René Clair showed a character who was complaining about the critical times they were going through, claiming instead that the past had been much gentler. While doing so, the decoration behind him would change: first it would be the twenties, then the eighteenth century, the Middle Ages, and so on, all the way back to the pre-historic age. Such feeling is understandable, for the future is uncertain and we yearn for certainty. Maybe this fear has some adaptive value. But the odds are *not* against human society.

The main flaw of the ecological crisis as a description of the human stance towards nature – since the concept is meaningless deprived of its social element, for nature as such would not disappear even if we do – is that it portrays as *critical* what is but a *normal* feature of socionatural relations. In the words of Ingolfur Blühdorn:

> So the concept of crisis implies either the return to the former state of normality or the transition to a new state of stability. But in the case of the natural environment and the societal relationship towards it, there is neither return nor stability. Particularly in a context where change, innovation and flexibility have been installed as the highest values to strive for, the term *environmental crisis* and its implications have become anachronistic (Blüdohrn 2004: 14).

Yet it is not so much that the term has become anachronistic, as that it has never been appropriate. Its usage entails a misunderstanding about the character of socionatural relations, whose normal state is precisely a state of crisis, namely, a constant process of reciprocal transformation and mutual adaptation between the social and the natural. A permanent crisis is hardly any crisis. Admittedly, different social values can make a difference in the way in which societies relate to their environments, but it is a difference that cannot really make a difference, since the overall process of human appropriation is not dependent on contingent social values: it is both an outcome of, and a precondition for, human evolution.

Suggesting that the ecological crisis may well be an imaginary one does *not* involve denying that the natural world is harmed by humanity in the course of that process, nor suggesting that this very process cannot be run differently. It can – although it is not so clear whether it *could* have been in the past. In any case, we should not mistake the loss of *natural forms* for a truly *environmental* deterioration. They are not the same thing, although environmentalism has traditionally claimed that the former involves automatically the latter. A moral position about the desirability of freezing socionatural relations – in a given state, at a given moment – is hence translated into a political category, that of the ecological crisis. It is not so much ironic, as melancholic, now that such a crisis is declared, that the human appropriation of nature has culminated with its disappearance: nature is now but the *human* environment. This point will be carefully developed later.

But it is this reality that environmentalism has yet to come to terms with: a context wherein the notion of an ecological crisis is meaningless.

As it happens, environmentalism has been traditionally bound to the notion of an ecological crisis, so that any challenge to the latter is taken as a challenge to the former. Furthermore, the identity of modern environmentalism has been mostly – albeit not totally – linked to the idea of an exceptional state of the socionatural relations, whose main expression is the damage done to the non-human world. If the crisis is questioned, becoming a *controversy* rather than a *condition*, environmentalism itself seems to weaken. More precisely, it happens to lose the absolute criteria that allow greens to place themselves on a higher moral or cognitive ground (Barry 2004: 180). In other words, environmentalism would be normalised, insofar as it would be placed *among* many others' conceptions of the good, not *apart* from them. The fierce journalistic debate that saluted the publication in the UK of Lomborg's *Skeptical Environmentalist* corroborates the central position occupied by the ecological crisis within environmentalism.

Does that mean that environmentalism cannot exist without ecological crisis? Not exactly. Maybe not the kind of environmentalism that has been mainstream since the rise of the green movement, but another environmentalism can exist – one that is viable and even necessary. In that regard, a realistic approach to sustainability provides a new basis for green politics, which is to become more pragmatic than radical, as well as liberal-pluralist rather than monist. It must not depend on the confirmation of the ecological crisis hypothesis; on the contrary, it must be grounded *beyond* survival. To some extent, this reinvention makes perfect sense as a late development of green thinking.

The Revolt against Classical Environmentalism

Despite the great amount of attention that human reflection has paid to nature throughout history, it was not until the emergence of the green movement at the end of the sixties that a properly political theory *about* nature was carefully assembled. Strictly speaking, green political theory is not constituted as such until this new kind of approach to nature – as a moral and political subject – is self-conscious and mature enough to venture into a more sophisticated realm, i.e. that of a truly original theory about nature as a political category. In that regard, an evolution of green thinking can be discerned, even though its different stages have not always been visible outside the movement itself. In fact, the main flaw of contemporary environmentalism lies in its inability to break convincingly with its own past, namely, the set of beliefs and proposals advanced by classical environmentalism. This must change – maybe it is changing already.

It is hard to underestimate the influence that early environmentalism has exerted on the constitution of the movement's identity, as well as on its theoretical development. A liability rather than an advantage, this heritage has hindered a much needed self-correction on the part of environmentalism, which despite its diversity has remained for the most part stubbornly loyal to a number of normative propositions only recently challenged within green thought itself. Such core propositions can be summarised as follows:

1. an ecological crisis is under way that threatens human survival and the preservation of the natural world;
2. the current socio-economic system, as the culmination of a long history of human alienation from nature, is to be blamed for it;
3. radical action is thus necessary, entailing a complete transformation of society, for it to become sustainable;
4. the subsequent sustainable society must incorporate a recognition of the intrinsic value of nature into its core values and practices, from which stems a human obligation of protection.

Of course, there are countless nuances, different intonations. Environmentalism is as diverse as any other political ideology. But mainstream environmentalism has retained a great deal of consistency from the outset, the *novelty* of its core foundations being precisely a reason to grant it a place between contemporary ideologies in its own right (see Dobson 2000a). Yet despite the powerful influence of classical environmentalism, such principles were not all written down from the very beginning. They are the outcome of a slow theoretical maturing, in the course of which environmentalism has gained as much complexity as, lately, true inner diversity. Paradoxically, if this late development opens up the opportunity to reframe environmentalism, it does so at the expense of those core principles – so that for many greens a renewed environmentalism would be no environmentalism at all.

To cut a long story short, once the founding fathers of environmentalism had sketched the main principles of the green movement, a more sophisticated exploration of them was due. Hence the next stage of green theory consisted in the unfolding, in many different directions, of the theoretical assumptions set by its early practitioners: the denunciation of the ecological crisis, a new conception of nature based upon the insights of ecology, a moral and spiritual concern for it, the wish for a radical transformation of society. Their exploration grows parallel to the development of both environmental sociology and philosophy.

This phase is at first an inner conversation about how a consistent green political theory can be devised. Afterwards, it reaches out to other theoretical perspectives, in what may be described as the exploration of the possibilities just opened by a new and powerful subject. From socialism and Marxism to anarchism, ranging through to feminism, classical and modern ideologies engage themselves with the green issues and proposals – a speculative dance after which they influence environmentalism as much as they are influenced by it. Some interesting hybrids are indeed born out of this encounter, such as eco-Marxism, eco-feminism or even social ecology, a personal mixture of anarchism and environmentalism (Pepper 1984, 1993a; Bookchin 1980, 1986; Plumwood 1993). The history of political theory is also re-interpreted in the key of environmentalism, and the same goes for history itself, wherein an explanation for the human alienation from nature can be traced (see Eckersley 1992; Pepper 1996; Glacken 1967; Passmore 1974; Plumwood 1993). In sum, Western political tradition *welcomes* environmentalism – like an ecosystem adapts itself to the presence of a new species.

At the same time, an irreducibly green strand of thought is patiently pursued, namely, the philosophical and ethical reflection on nature itself. The pioneers of so-called *deep ecology* had invoked since at least 1972 the need for a more spiritual contact with nature, hence pointing to human consciousness as a major factor of social change (see Naess 1989; Devall and Sessions 1985). By identifying the values upon which human obligation towards nature can be defended, among them recognition of its intrinsic value and its inclusion in an expanded moral community, environmental philosophy at large provides an ethical foundation for green political theory (see O'Neill et al. 2001). It is to be noted that animal rights are a relevant part of this moral extensionism, opening up the possibility of an entitlement for the environment at large (see Stone 2010 [1972]; Regan 2004 [1983]; Benton 1993a, 1993b; Singer 1976). Such metaphysical foundation, albeit one which is against traditional metaphysics, has not always benefited green politics, resting as it does on a contestable conception of nature. However, environmental sociology, as well as a number of philosophers, be they environmental or not, have come to *problematise* that view of nature, opening the door to a more sophisticated account of socionatural relations (Dickens 1992; Soper 1995; Cronon 1996; McNaghten and Urry 1998; Newton 2007).

On the other hand, a crucial political notion made its way to the public debate over the course of these years, after two decades of a somewhat subterranean existence. In 1992, the Bruntland Report on sustainable development was published, introducing an idea that immediately dominated the environmental debate (WCED 1987). An intense discussion was opened within environmentalism about the concept and its implications, among them the issue of inter- and intra-generational social justice (see Redclift 1987; Lelé 1991; Dobson 1998, 1999; De-Shalit 1995; Jacobs 1999). Global warming has just come to complicate matters, by demanding an even more comprehensive conception of sustainability (see Dessler and Parson 2006). Still, global warming can be conceptually absorbed by sustainability. It is environmentalism, rather than sustainability, that may be disturbed by the appearance of the mother of all environmental problems.

The consolidation of green *political* theory took place throughout the nineties, when a number of authors tried to articulate politically the main principles of environmentalism (see Dobson 1990; Eckersley 1992; Merchant 1992; Dobson and Lucardie 1993; Hayward 1995). Green political theory emerged as a radical ideology, a challenge to liberal democracy rooted in an ecocentric philosophy, whose main goal was the achievement of a sustainable society very often depicted in the guise of utopia. However, the naturalist foundation of green political theory led to a paradoxical depoliticisation of politics – insofar as the contingency of the latter is discarded on account of a rigid metaphysic. Attempts were made to solve these contradictions, either exploring the kind of justification underlying the application of green policies, or reflecting upon the difficult relationship between an environmentalism so conceived and liberal democracy (see Goodin 1992; Norton 1991; Doherty and De Geus 1996; Lafferty and Meadowcroft 1996).

Yet a more interesting move is taking place now. In recent years, environmentalism has started to question itself, deploying a reflexive stance that is leading it in new directions. Starting in the mid-nineties, some authors began to question tenets such as naturalism or radical democracy as being the right green foundations, thus leading the path for an immanent critique of their own concepts, ideas, and value commitments (see Hayward 1998; Barry 1999; Torgerson 1999; Meyer 2001; Blühdorn 2004a; Humphrey 2002). In other words, green theory started to question green ideology. As if Kirkman's advice had been followed:

> let those of us who identify ourselves as environmentalists examine our own assumptions and practices with at least the same degree of critical candor we bring to bear on the assumptions and practices of the status quo (Kirkman 2002: 4).

Nevertheless, this shift can also be seen less as a challenge than as a move forward. In the preface of the third edition of his canonical *Green Political Thought*, Andrew Dobson suggests that, instead of insisting on his own ideological principles, environmentalism is now reflecting on traditional political concepts, ranging from justice to citizenship, and also *greening* them (Dobson 2000a: ix). On the one hand, this exploration produced an abundant literature on the plausibility of a particular green democratic model, mostly a post-liberal one with a strong deliberative accent (see Mathews 1995a; De-Shalit 2000; Minteer and Pepperman Taylor 2002; Hinchman 2004). On the other, a formerly unthinkable dialogue with liberalism was opened (see Wissenburg 1998; Vincent 1998; Levy and Wissenburg 2004a; Hailwood 2004). However, environmentalism may have changed course as a result:

> I argue that, as the environment has become an increasingly mainstream issue, so the centre of gravity in environmental politics has shifted from a *radical* rejection of contemporary society and a relatively *narrow* concern with ecological issues, to a *reformist* acceptance of capitalist liberal democracy accompanied by a *broader* social justice agenda (Carter 2007: 9).

Needless to say, the weakening of green radicalism can also be seen as a capitulation, a "post-ecologist turn" in the face of liberal environmental policies that are but "simulative politics" (Blühdorn 2004a: 260, 267). Nonetheless, the reflective moment of green political theory can also be interpreted as a revolt against classical environmentalism. Rethinking environmentalism would then be partly fuelled by a *suspicion* about the suitability of the classical green foundations. It has not just created the opportunity to de-radicalise environmentalism, turning green ideology into environmental politics, but it has also provided the chance to reframe it in a more realistic and appealing way.

Environmentalism meets Global Warming

Is the rise of climate change a triumph for environmentalism, or a peculiar curse in disguise, marking the beginning of its end? On the face of it, this looks like an absurd question. The green movement has long campaigned to stop global warming, and the success of this mobilisation is apparent: fear of a climate dystopia has made its way into contemporary culture and governments have incorporated it into their agendas. A modest change so far, maybe, but a significant one. The private ecologist must rejoice at the influence finally exerted by the public ecologist, expecting that the implementation of a climate change policy will eventually lead to an ecocentric re-orientation of the socionatural relations. In sum, it is hard to see how climate change can be detrimental to environmentalism.

Yet it can. Despite all the greens' hopes about the wrong reasons paving the way for the right outcomes, it is far from clear that the outgoing social action will end up in the kind of sustainability traditionally defended by them. The advent of global warming has already transformed environmentalism by changing the world around it. Sustainability has acquired a new meaning: instead of saving nature from humanity, humanity is to be saved from nature. In other words:

> When concern about climate change went mainstream all over the world in 2007, Greens everywhere felt vindicated. (…) Wrong. The long-evolved Green agenda is suddenly outdated – too negative, too tradition-bound, too specialized, too politically one-sided for the scale of the climate problem. Far from taking a new dominant role, environmentalists risk being marginalized more than ever, with many of their deep goals and well-honed strategies irrelevant to the new tasks (Brand 2009: 209).

The ecocentric values, or at least the anti-anthropocentric stance, that have always been the green features *par excellence*, seem now obsolete, as if the public ecologist had accidentally outplayed his private incarnation. That is certainly a bitter irony, but it may also reveal something deep about the classical green discourse – namely, that it does not belong to this world. Not anymore. Or not just yet.

An unofficial split within environmentalism is taking place, separating those who aim to push environmentalism in a new direction and those who believe the alternative to be a detour that happens to be not green at all. The resulting landscape is endearingly varied. Some advocate a technologically rejuvenated environmentalism (Mol 2008; Brand 2009; Greer 2009; Tomlinson 2010), while others insist on pursuing ecocentric goals going back to nature (Haila and Dyke 2006; Sale 2006). There are those for whom a correction of capitalist markets – without overthrowing them – is the most realistic path to sustainability (Porritt 2005;

Spen Stokes 2009), whereas for others it can only be achieved after capitalism is dismantled (Albert 2006; Patterson 2009). The advocates of a more reformist environmentalism (Carter 2007; Humphrey 2003) clash against those who see it as a radical ideology which has to remain so (Blühdorn 2007; Derber 2010), whereas others advocate the overcoming of pessimism and the adoption of a more positive narrative for environmentalism and its goals (Schmidt 2005; Wapner 2010; Princen 2010; Schor 2010). A divide is also perceptible in the field of sustainability studies, wherein defenders of a strong sustainability demand a complete breakup with the current social organisation, whereas supporters of a weak sustainability prefer a more moderate approach and believe in greening the current social model, a view otherwise expressed in the theory of ecological modernisation (Ott and Döring 2004; Schmidt 2005; Neumayer 2010; Mol et al. 2009).

There are a number of ways to interpret such a division, as far as the identity of environmentalism is concerned: deviationism *versus* authenticity, pragmatism *versus* normativism, reformism *versus* radicalism. None of them, however, is complete enough. Pragmatists trying to put political outcomes before principles have principles too; in the end, radicals may achieve less revolutionary outcomes than reformists; and so on. Maybe the most accurate and simplest distinction is that between those trying to move green politics beyond its traditional – ecocentric *or* anti-capitalistic – foundation, without predetermining the outcome of such theoretical venture, and those who resist such attempts defending environmentalism as a nature-centred and hence more radical ideology. The latter are still a majority within the green camp.

However, I would like to suggest that such split can be simplified by presenting four *successive oppositions* that give form to an alternative set of ideas about sustainable society and hence green politics. Sorting out these criteria helps us to understand the crossroads environmentalism is at nowadays. Besides, they allow us to shed light on the relationship between environmentalism and other political ideologies and concepts – from democracy to liberalism.

(a) *Recognition of nature* versus *end of nature*. This cleavage separates those who still believe in nature from those who claim that it does not exist anymore. The former maintain that nature has some kind of intrinsic value that deserves a social recognition, which in turn should lead to a greater degree of protection in order to grant it its autonomy. Conversely, the latter suggest that nature has been humanly appropriated in a deep sense in several ways, so that it is meaningless to speak of it as a separated realm. Nature is thus to be recognised as a part of society – a part of the human environment.

(b) *Highly restricted* versus *widely accepted natural substitutability.* Opposing those who believe nature to be either morally or technically unsubstitutable by human-made artifacts, against those for whom that substitution is both technically feasible and morally acceptable. This substitution does not need to be total to be controversial – a significantly high degree of technical substitution suffices to raise the question about its desirability. Among the former, some may believe substitution to be technically possible but morally unacceptable. It is to be expected that those who claim that nature's value must be recognised will oppose substitutability, whereas those who suggest that nature has been dissolved into society will support a higher degree of substitution.

(c) *Closed* versus *open sustainability*. Although the opposition between *strong* and *weak* versions of sustainability is both meaningful and useful, it does not express all that there is to know about sustainability. Above all, sustainability's relationship to democracy may be overshadowed if we only look at the substitutability question just described. If we distinguish instead between *closed* and *open* sustainabilities, substitutability remains crucial, but the socio-political dimension of sustainability is also highlighted. Hence a restricted substitutability will lead to a *closed* model of sustainability, wherein planification plays a significant role, insofar as the shape of the sustainable society is sketched in advance in order to serve a moral recognition of nature previously decided upon. On the contrary, an *open* version of sustainability is based on a high degree of natural substitutability and does not determine in advance the content of a sustainable society, which remains uncertain and will eventually emerge as the outcome of a liberal-pluralist society, wherein a permanent controversy about the good life is *discussed* in the public sphere and *played* in the practical realm. However, accepting a high degree of substitutability *prima facie* is not equal to implementing it – that will depend on the social consensus about the issue.

(d) *Conservative* versus *transformative environmentalism*. The previous oppositions pave the way for a more general appreciation of environmentalism's character, namely, the fundamental orientation of its core values and policy proposals. On the one hand, those who promote a strong recognition of nature's value, defending the need to restrict its substitutability through a closed version of the sustainable society, can be said to stand for a *conservative* environmentalism, namely, one whose purpose is scaling back society for the benefit of nature. This is also the attempt to conserve a certain understanding of the socionatural relations, as opposed to a more open and contingent view of the latter. On the other hand, a *transformative* environmentalism does not restrict the natural substitutability in light of the irreversible interweaving of nature and society, believing in the human ability to re-orientate the socionatural relations in a sustainable, but also human-enhancing, way. As noted, the degree to which natural forms are protected will depend on the social consensus about it. Needless to say, a zero-growth sustainable society would also be transformative, but its foundations are, strictly speaking, conservative: a predictable future where human aspirations are limited, so that nature can flourish.

The battle for the soul of environmentalism revolves around these oppositions. They will be explored throughout this book. It is noteworthy that they focus mainly on the green stance on *nature*, as the central foundation of any version of environmentalism. The remaining features of each of these variations will stem from that core proposition – be it the conception of sustainability or the relationship with liberal democracy. Thus the particular *difference* of environmentalism as a political ideology is respected, no matter how relevant some other goals, such as social justice, can become in the green agenda.

Be that as it may, the rise of global warming has provided the opportunity to reframe environmentalism. There is not, and there should not be, such thing as a *true* environmentalism: the re-organisation of socionatural relations is too important an issue to be left to the hands of orthodox greens. It is obvious that it would have been impossible to reach this point of cultural maturation regarding the environment without the pressure exerted by a deep green position. Paradoxical as it may be, a *radical* discourse is necessary to mobilise society around a *moderate* agenda: a bitter victory if there ever was one. However, if greens are to play a significant role in the forthcoming environmental debate, they have to move forward.

Reframing Environmentalism

This book advocates a shift for environmentalism. It suggests that green politics should be reframed in the direction of a more realistic position towards the socionatural relations. This entails a view of sustainability as a means to refine the human domination of nature. Instead of a rhetoric of doom, environmentalism should advance a more inspiring narrative for humanity, one which, being sympathetic to its past record as a species, encourages a general improvement of society *through* sustainability. In turn, such sustainable society can only be reached within the normative and institutional frame provided by liberal democracy – albeit properly corrected. The outcome is a post-natural environmentalism which goes liberal and becomes more realistic while remaining green.

Rethinking environmentalism in this way involves a *nuanced* refusal of naturalism. Nature has played too big a role in the articulation of green political theory, especially since the view of nature that has been dominant so far in green philosophy is significantly flawed. The idea that we can go *back* to nature, closing the gap opened by a slow but constant process of alienation from it, belongs to the utopian realm. A sustainable society centred around the restoration of human harmony with nature is, in fact, a projection of a retrospective utopia: that of a nature that exists separated from society and outside history. Proust once noted that each paradise is a paradise lost. And indeed such pastoral landscape is not located in the past, because it never existed, so any attempt to reproduce this Arcadian dream is doomed to fail. Arguably, nature does not belong to the realm of ideas, but to the realm of desire (Rosset 1974: 28).

Thus the green view of nature constitutes a *mythology*, in the sense proposed by Roland Barthes (1993). It is not a myth that hides reality, but rather one that distorts. When nature is portrayed as a universal and ahistoric entity, it is also disembodied and reduced to an abstraction. The historic and social condition of the natural world – its social history – is thus sidelined. With the passing of time, nature has come to be a historic and social construction. However, that process is naturalised whenever the non-human world is portrayed as an independent entity isolated from society. On the contrary, the human appropriation of nature has led to a growing interweaving of nature and society that culminates with the disappearance of any separation between them, turning nature into human environment.

In this regard, a successful reframing of green politics includes the reformulation of a classic green theme: the human domination of nature. The traditional approach to the human-nature relationship involves a plain accusation to our species, namely, that it has subjugated the natural world throughout history, a dominion deemed morally and ecologically wrong. Yet such practices of domination can also be understood as practices of adaptation, developed well before an ecological consciousness could even emerge. Domination is, in fact, a precondition for care. Instead of levelling an accusation to humanity on the grounds of a historic mistreatment of nature, we can reframe domination as the adaptive answer of human species before a greater ecological consciousness could be acquired *through* domination itself. A narrative of *sympathy* towards the human species, rather than one of *guilt*, may then be fairer – and more useful. Humans were not simply co-existing with a peaceful natural world; most of the time, a not too well educated humanity was – still is – fighting her way out of a threatening environment. This is not conservatism, nor historical determinism: it is the way things *were*. They, us, knew no better.

A shift from blame to sympathy serves a transformative environmentalism well, insofar as the quest for sustainability can therefore be reframed as a process of *refinement*: a refinement of the socionatural relations and hence a refinement of the human species as well. Such a reflexive re-organisation of our relationship with the natural world involves a reflexive domination. But an improved domination of this sort in turn requires material progress and a further betterment of the human condition, which are to be obtained through human ingenuity and technological innovation. Because a pacification of the socionatural relations cannot be achieved by dismantling post-industrial societies, not even by convincing their inhabitants that it is in their interest to do so. That is what sustainability after the end of nature is about: a human politics for the refinement of the socionatural relations. Thus humanity is reconciled with nature *through* sustainability *within* the environment. A more inspirational frame for environmentalism – from doomsayer to civilisation-enhancer – thus emerges.

Of course, the question remains of *how* the sustainable society is to be achieved. Environmentalism has traditionally oscillated between two conflicting positions: a temporary suspension of democracy in the name of ecological efficiency and a radicalisation of democracy in the name of popular wisdom. Only lately has it come to terms with a more liberal stance promoting reform instead of utopia. However, green politics are to be reconsidered, together with sustainability, in the light of the end of nature. Such philosophical framework helps us to understand both sustainability and the relationship between sustainability, democracy and liberalism in a new way, namely, as mutually reinforcing elements rather than conflicting ones. If only because a green agenda, no matter how urgent the task may seem, cannot be *imposed* upon society if it is missing a cultural consensus about it. Thus sustainable society can only be a liberal green society – a society, on the other hand, which is arguably in motion already. Therefore, the politics of sustainability must be more liberal than democratic, i.e. they will be democratic by different means – including e-government, collaborative democracy and the market. Herein lies the future of a renewed green politics.

The resulting environmentalism is not to be conceived of as a *moral* doctrine committed to the protection of the natural world, but rather as a *political* theory whose goal is the achievement of sustainability. It is the reflexive balance of socionatural relations which is at stake – a balance that does not necessarily revolve around the protection of natural forms, rather one which is based on the search for the preferred *interaction* between society and the natural systems. Nevertheless, there is no restriction on the protection of those forms. A realistic environmentalism is not *against* nature, only against self-delusion *about* nature.

In sum, green politics must not tie itself to the arguments proposed by classical environmentalism. It can be fruitfully reinvented, since there is no single definition for it. Still, a realistic environmentalism should be seen as a continuation, rather than a radical departure, for green theory. By reframing green politics, a more realistic and pragmatic stance can provide environmentalism with a renewed influence in the sustainability debate – an influence that otherwise may well lose. So, recalling the old story mentioned at the beginning of this book, it is the time to realise that, if we do not want to get lost in our way to the sustainable society, the very wish to go there demands a new start – from somewhere else. A change of perspective that concerns, first and foremost, environmentalism itself.

Real Green: An Overview

This book is organised as follows. The first part is devoted to offering a realistic account of the relationship between humanity – thus society – and nature. Such account must be able to explain why nature has morphed into human environment and what the consequences of it are. This view is opposed to the conception of nature that has become dominant in the Western imagination, no matter how much it is refuted in our everyday exchanges with the natural world. Environmentalism portrays nature as an ahistoric, autonomous and universal entity. But it is nothing of the sort. The first chapter starts by explaining how society is actually *within* nature, namely, how nature is a social construct. Not a linguistic or cultural construction, but a literal one: we have re-made nature and, in doing so, we have re-made ourselves. Such social quality of nature helps us to understand classical green issues in a new light, such as the existence of natural limits to human activity or the claim that nature has an intrinsic value.

Although society is within nature, humans are also undeniably a part of it: we are an animal species, albeit a particular one. That means that nature is within society too. However, although environmentalism has always underlined humanity's naturalness in order to suppress the dualism between nature and culture, we should think twice before that border is removed. We do not fancy that we are separated from nature; we are truly separated from it. The second chapter will discuss the debate about human and animals, as well as claiming against the naturalistic reduction of the human condition. Humans are indeed exceptional and are to be considered as such. In fact, it is argued, nature's protection is not well served by making humans closer to it, but rather by fostering a sense of wonder and distance towards the natural world. The apparition of a new form of naturalism, i.e. the geneticist reading of human nature, will also be discussed, insofar as it can truly affect our understanding of human nature.

The next chapter brings the previous ones to their logical conclusion, claiming that nature does not exist anymore. In connection with it, the relation between nature and artifact is considered. Why do we value more the natural, i.e. what was there, rather than the artifacts we have been able to achieve with our ingenuity? The domination and humanisation of nature is the next logical step in this philosophical itinerary. A rebuttal of the green critique of human domination must be performed, in order to understand such dominion as an essential human trait. This understanding allows us to rethink dominion as a reflexive control of our interactions with nature. Sustainability is then to be understood as a legitimate project for the enlightened domination of the latter. The rest is self-delusion. A subsequent distinction is then to be made between nature and the environment.

Once the philosophical grounds have been made clear, the book must move to a different territory, that of the means to make our society sustainable. Such is the topic of the second part of the book. The fourth chapter is devoted to explore the concept of sustainability itself. Its ambiguity and complexity are underlined, as well as its normative character. It will be argued that there is no single sustainability, but many possible sustainabilities. However, a worrisome trait of this principle will be discussed; namely, an escathological tendency to put it once again into the future, as though a sustainable society would be an update of the old Marxian classless society. For several reasons, the mythology of sustainability is to be replaced by a realistic view of it, including the realisation that sustainability is already under way. On the other hand, sustainability has been affected by the sudden irruption of climate change into the public debate. After all, the former must indeed comprise the latter. It would seem that a climatic threat refutes the idea that nature *is* ended or can *be* dominated, but the contrary will be suggested: climate change confirms rather than contests the need for a post-natural frame of mind when dealing with sustainability.

A key question remains about the relationship between sustainability and democracy. Is sustainability compatible with democracy? What kinds of sustainable society are possible? Does sustainability determine or condition social organisation? To answer these questions, the fifth chapter starts by discussing the conflict between environmentalism and democracy, i.e. the authoritarian temptation within green thought. If greens cling to a naturalistic view of nature and to an apocalyptic reading of history, the latter will not be just a curious episode in environmentalism's history, but will remain a well-defined possibility for it. It follows from here that a clear distinction can be made between two kinds of sustainability: a technocratic sustainability that is incompatible with democracy; and a normative sustainability that is radically open and hence plausibly democratic. The latter can also be either radical-democratic or liberal, a distinction that will be recovered and explored further at the end of the book.

Liberalism and liberal democracy have for a long time been the targets of green criticism. To greens, liberal values and institutions are partly responsible for an ecocidal culture and hence a hindrance to sustainability. Hence the classical defence of different forms of radical democracy and community as an antidote to liberal-capitalism. As of lately, a part of environmentalism has been more inclined to accept liberal reformism, although such movement is far from being unanimous within green thinking. The third part of the book is devoted to demonstrate that the future of green politics must be liberal, i.e. that the sustainable society must be a green liberal society. To such end, the sixth chapter outlines the conflict between environmentalism and democracy and defends the possibility – even the reality – of a green liberalism, grounded on the accounts of both socio-natural relations and sustainability described in the first half of the book.

But what is a green liberal society like? How does it differ from the classical green view of a sustainable society? How is the liberal green society to be *politically* articulated? Although this question allows for a number of approaches, given the variety of institutions that may be potentially greened in order to accommodate ecological concerns in liberal democracy, ranging from political representation and rights to the state and the principle of autonomy, the discussion here will revolve around two closely related issues: the democratisation of sustainability and the notion of ecological citizenship. After all, given that an open conception of sustainability is to be realised in a democratic society, it remains to be clarified how such an open principle is to be implemented on a practical level and by whom. In other words, how is the social change required by sustainability to be achieved? And what is the role of democracy and citizen participation in that process?

Whereas green thinking has usually stressed the blessings of democratisation, the seventh chapter will try to show that, on the contrary, we should stress the limits of democratisation when dealing with sustainability. To be more precise, it is direct democracy that has to be limited, whilst a wider conception of democratic participation is to be defended – including the participation through the market, the informal deliberation in the public sphere and the communication struggle that social movements engage in for a definition of the good life. An enormous and complex system of debate thus emerges, in which governments, citizens and collective actors simultaneously *act* and *communicate* around the key issue of sustainability. It is during such debate that sustainability advances through the modernisation of society and the refinement of the socio-natural relationship.

As a particular exploration of this problem, the seventh chapter focuses solely on the green redefinition of citizenship and its resulting hybrid: ecological citizenship. Not only because of its current relevance in the environmental debate, but also in view of its undeniable usefulness. There can be no such thing as a sustainable society without committed, or at least, compliant, citizens. After all, citizenship is an institution located between the formal political organisation and the informal realm of civic action, as well as between the public and the private, while simultaneously related to the market. It is thus able to reflect the nature of a liberal social order, while serving simultaneously as a tool for its transformation. Besides, it is directly related to people's motivation to behave in a sustainable way. In sum, it helps us to reflect on the relationship between sustainability and liberal democracy.

The concluding section of the book deals with the future of green politics. It does so by discussing the hypothesis of the end of environmentalism, i.e. the idea that environmentalism has dissolved into liberal democracy once the aim of sustainability has gained ground in the contemporary public agenda. Likewise, a proposal is made for a renewed green politics, one that is able to abandon its old tenets and finds a new discourse which is less utopian and more liberal than the current one: a real rather than utopian green.

PART I
Nature and Society

Chapter 1
Society within Nature

Neither Here nor There?

The quarrel about the human ability to know the world has been an important part of philosophy from its very beginning. It is not surprising that such importance increases when it is the relationship between society and the natural world that is dealt with. In fact, there is an obvious link between the question about nature's nature and the principle of sustainability, so that a seemingly philosophical inquiry ends up being essential to *political* decision. However, such debate has so far been centred around an opposition – between realism and constructivism – which should be put into question. The reason is its inability to properly explain a *fact* which ought to be the standpoint of any debate around sustainability: the transformation of nature into human environment. Therefore, in order to make sense of the transformative interaction between society and nature, neither realism nor constructivism are as such convincing. A realistic turn in the constructivist account of nature is thus advisable.

But first thing's first. Grounded in different theories of knowledge, realism and constructivism defend two opposed conceptions of nature, and hence our ability to gain access to it. According to the realist position, nature is an objective, autonomous, independent entity. It can be known, but is firmly located beyond the social realm. On the contrary, the constructivist position refers to "nature" as the final outcome of a process of social construction through language and culture. It is something we can only know through society itself. But whereas radical constructivism denies that nature has an independent ontology, moderate constructivism claims that we should not even worry about such ontology, since we cannot grasp it. Needless to say, each position is related to a different understanding of the socionatural relations and henceforth helps to support different moral orientations towards nature, as well as disparate models of sustainability.

Unsurprisingly, constructivism came early under attack from a large part of environmentalism, which accused the former of lacking enough moral compromise with the *defence* of nature (Dunlap and Catton 1994; Benton 1994; Martell 1994). The underlying rationale for this accusation was that whoever embraces a constructivist position is also adopting an epistemological neutrality incompatible with a green stance. More recently, this sort of argument has echoed in the assertion that the recognition of nature's social dimension has hindered the development of a transformative environmentalism, insofar as it has downplayed the conflict between the current social model and a truly sustainable society (Blühdorn and Welsh 2007: 192). In the words of Valerie Plumwood: "The deep contemporary suspicion and skepticism about the concept and term 'nature' may play some role in the contemporary indifference to the destruction and decline of the natural world around us" (Plumwood 2005: 27). In sum, now and then constructivism is seen as a declaration of war against nature (Lease 1995: 4; Rogers 2009: 1). All of which comes to show that this debate is far from being mere philosophical charlatanry. After all, if the natural world had *not* an independent value, political environmentalism should not reason *as though* it had such value; and vice versa. That nature is either *out there* or *just here* thus has a formidable importance.

Henceforth, despite its internal diversity, environmentalism has mostly defended a view of nature that falls into the realistic side of this divide. Therefore nature is understood as a fairly independent entity which has to be separated from society, so that we can recognise its moral value and make the political arrangements that are necessary to protect it. Historically, human society has developed itself *against* nature. Now the time is ripe for a conceptual and practical disentanglement between them that facilitates not just the recognition, but also the protection of, its autonomy. Both nature's value and autonomy are intrinsic to it and thus independent of human evaluations. Nature does not exist for *us*, but for itself: it is self-generated and self-sustained, i.e. it is autonomous. This also means that nature is a universal entity, whose essence remains untouched by human history. The moral recognition of the philosophical autonomy of nature is thus to be granted in the practical realm via a – politically organised – human withdrawal from it.

Such view of nature stems from a peculiar combination of science and intuitionism. Although it is based on the insights of ecology, the account of nature so obtained inspires a new apprehension of the natural world which goes beyond the scientific realm. On the one hand, nature is perceived as an intricate web of connections, a complex and enormous organism whose parts cannot properly be separated from the whole. Such wholeness cannot be grasped in terms of competition or survival, since it is a living form of cooperation, a holistic order that reveals a barely Darwinian picture. As Murray Bookchin puts it: "Life is active, interactive, procreative, relational, and contextual" (Bookchin 2001: 67). Yet, on the other hand, the same science that reveals nature's complexity is unable to reduce it completely, as Aldo Leopold (1987: 33) himself once celebrated. This means that nature can also be experienced beyond knowledge, i.e. in the spiritual realm. The individual self can close the gap between nature and culture rediscovering its own naturalness, a sense of belonging long hindered by the disenchantment of nature. Human identification with nature – coming *back* to it – is meant to serve as an emotional basis for the greening of ethics and the political construction of a sustainable society.

However, paradoxically enough, a supposedly realistic account of nature ends up being not realistic at all. By pulling human society out of nature, environmentalism offers a too abstract depiction of the latter that fails to explain the interaction between society and nature. Hence I will suggest the need to re-think the opposition between realism and constructivism, in order to plausibly explain the state of socionatural relations. Hence a realist turn in social constructivism is to combine the virtues of the latter with the inescapable material dimension of the human-nature interaction. We do not only construct nature's concept, but also nature's reality. In other words, the social construction of nature is not just an idea, but the material outcome of a relational process of co-evolution which has been going on for millennia. Its final result is the transformation of nature into human environment.

The Social Construction of Nature

Despite the green attempts to propose something akin to a natural definition of nature, the fact that the latter is humanly appropriated and socially reformulated must be the starting point for any reflection on the socionatural relationship. This is not to deny the physicality of the natural world, rather to point out that the latter enjoys only a relative independence from society. Nature is rather what we have constituted as nature through a process of social construction. But let us explain this.

The constructivist account of nature is but an application to the latter of the wider paradigm of social constructivism. Its roots lie arguably in Marxist sociology of knowledge, and its influence, especially after the seminal work of Berger and Luckman (1991), has constantly grown – as can be seen in psychology and in the multidisciplinary development of frame theory. Both the so-called linguistic turn and the post-modern defence of relativism have helped to spread it.

Its starting point is fairly simple: reality cannot be objectively and directly seized. It is always observed, classified and even experienced through social categories. The power of the mediation is such that these categories end up *creating* that very reality, hence revealed as *mostly* a social creation. Rather than *a* reality, there exist several *realities*, which are relative to different linguistic or cultural points of view. This is precisely what an objectivist position fails to take into account – the degree to which reality is never reality *as it is*, but a culturally and socially filtered entity. Thus constructivism tries to identify and explain the devices upon which each society or culture *builds* its world image. Unsurprisingly, language plays a central role in this account, insofar as it is the foremost mediation between humanity and the world. And that is why discourse becomes a key factor in the so-called social creation of reality. Therefore, any attempt to naturalise the social, i.e. to say that something is natural instead of constructed, is seen as an attempt to legitimise social conventions which, rooted in culture and history, serve an ideological purpose. On the contrary: everything is constructed, everything is language.

So, if reality at large is a social construction, nature is another one. Because the natural world cannot escape from social mediation: the truth about nature would then be its social truth. Nature is not a timeless essence, nor can be conceived of as something separated from human existence. Hence the *universal* nature invoked by environmentalism can only be attained in a very stylised, abstract way –

as the very ground on which everything is based upon. But this sense is too wide to be meaningful, especially since it cannot be *protected*: if we understand nature, to quote John Stuart Mill, as "all the powers existing in either the outer or the inner world and everything which takes place by means of those powers" (Stuart Mill 1998: 8), well, nature is going to exist as long as something at all exists. Furthermore, this social condition also means that there is no *single* universal nature, because different contexts, cultures, social positions and historical moments will produce disparate visions of nature (Macnaghten and Urry 1998). Instead of a unique nature, different natures co-exist, each of them consisting of different socio-cultural processes from which they cannot be freed.

However, the constructivist account of nature is not a mere application to nature of social constructivism writ large – there is more to it than that. Above all, the former is related to the emergence of philosophical naturalism and the subsequent opposition between nature and culture. Reacting against the biologist thinking that succeeded after the coming of Darwinism, the founding fathers of sociology made a double statement: on the one hand, the social is (relatively) autonomous from the natural; on the other, humanity posseses qualities not reducible to the sum of its natural features. Hence the fact that the human species is the outcome of natural processes does not mean that everything we do is *natural* (Moriarty 2007). Unless we understand such statement in the most trivial sense! Therefore, nature does not exist independently from humanity, nor is humanity just a part of it. Such is the conclusion that serves as a basis for constructivism. It can be seen as a reaction against the view that social evolution is only a *continuation* of natural processes. Consequently, the natural constitution of society is to be replaced with the social construction of nature.

Nevertheless, despite some common features, there is no single form of constructivism. On the contrary, it is already customary to distinguish between two varieties: radical and moderate (see Sismondo 1993).

1. On the one hand, *ontological or radical constructivism* points to the very essence of nature, to nature in a metaphysical way. It denies the independent existence of a natural world, since the latter is said to live only within social discourses. The natural world is but another social category. As such, this version of constructivism is a form of anti-foundationalism. It claims that the outer reality is socially set up and hence is not external at all. Nature itself is inaccessible outside language. In order to know nature, we must get to know our construction of it.
2. On the other hand, *epistemological or moderate constructivism* reduces such contextual dependence to nature's knowledge. Although nature is an actual entity, which enjoys a full existence on its own, our access to it is unavoidably subjected to several mediations and cultural filters that ultimately prevent us from knowing it directly. The natural world cannot be seized nor known outside these social mediations, so that there cannot be any universal nature, but instead disparate constructions – socially and

historically bounded – of it. Our singular seizure of the existing natural world is what we call nature. In Kantian terms (Kant 1986), nature is the *phenomenon* that we are able to describe through culture; the natural world is the *noumenon* that remains beyond our cognitive powers: the object is real but unattainable, the knowledge is constructed.

To be sure, green sociologists have mostly attacked the ontological or radical version of constructivism (Dickens 1996: 74; Martell 1994: 131). On the face of it, its claims are untenable and seemingly easy to refute. No matter how many ways there exist of constructing an animal *culturally*, none of those discourses have anything to do with the way in which an animal is constituted *organically*! Fair enough. However, maybe radical constructivism is not trying to say anything of the sort. Perhaps it is suggesting that we should suspend our judgements about nature's ontology, since our perception of the latter depends on our cultural representations – representations which happen to precede our perception. The *idea* of the animal is *the* animal for us. Since we cannot be sure that our judgements on reality are equivalent to reality itself, such judgements are maybe useless (Smith 1999). But it is dubious nonetheless that this approach can be useful for studying the socionatural interaction.

The truth is that sociological practice rarely endorses such radical version of constructivism. On the contrary, moderate constructivism is dominant in the field (Burningham and Cooper 1999: 303). Its purpose is to unravel the ways in which *social* reality is constructed, paying attention to the social processes that have an influence on the production of science and knowledge: "Whatever the *external* realities, for constructionist environmental sociologists the focus is on the question of how contemporary societies frame and process their *knowledge* about this *external* world" (Blühdorn 2000: 48). In this regard, Ted Benton (2000) aptly distinguished once between the object's reality and the knowledge's construction: natural processes exist outside humanity, but our knowledge of them is a social construction. The sociology of scientific knowledge will assume the task of unveiling the reasons that explain why the outcome of a social process leads to different beliefs and representations – i.e. to a given social construction – in each case.

Be that as it may, a conclusion is normally reached that accepts the epistemological and refuses the ontological version of constructivism. A reasonable *tertium genus* is thus agreed: we acknowledge both reality and the socially-bounded character of our knowledge of it (Peterson 1999). Therefore, nature is not considered a social epiphenomenon, but an independent entity that can only be known by us in a limited, indirect way. An additional advantage of this conclusion is that it leaves the door open to other ways of exploring nature – be they alternative scientific discourses or the mixture of aesthetic and spiritual search advocated by deep ecology.

In Defence of a Realist Constructivism

From a green standpoint, it makes sense to prefer a moderate version of constructivism, as a sort of lesser evil, namely, a way to maintain a moral compromise with the protection of nature. Sustainability cannot be based on an ontological constructivism, but essentialist realism is not so easy to accept either. Hence the middle ground: nature is neither fully autonomous, nor fully dependent on society. It sounds reasonable.

However, it is time to move beyond this binary code. The latter forces us to see nature *either* as the set of our material existential conditions *or* as a series of culturally generated symbols, but we should be ready to see it as *both*. The natural world is not a universal entity, but neither is just a concept. An exclusively realistic or constructivist explanation of the socionatural interaction eliminates one of the two parts of the nature-society *relation*. It is precisely the lack of attention to the material dimension of the latter that seriously hinders the usefulness of both moderate constructivism and realism. The point is that, if we distinguish between the natural world's existence and our social knowledge of it, we underestimate the crucial importance of human *practices*, that is, the actual physical interaction between society and nature. In this vein, Michael Redclift has pointed out how "the literature on nature and social construction has given considerable attention to the way in which we construct our view of nature, but much less attention to the way in which nature influences and transforms us" (Redclift 2006: viii), whereas Philip Sutton demands "ways of connecting the social and the natural within a single framework that would enable a new research programme for sociologists interested in the environment" (Sutton 2004: 68).

Henceforth, together with the ontological and epistemological versions, a *material* constructivism is to be acknowledged, in order to give a realist turn to this debate. It is by doing so that we can explain the actual transformation of nature into human environment. Because the construction takes place here as such: as a human intervention upon nature. In this sense, nature is physically re-constructed by humanity, in the very act of its appropriation, through the mixing of society and nature. In the words of Stephen Vogel: "*Construction* must thus be understood literally, as referring to the physical practices of transformation that can always be discerned to have been at work in the environing world we inhabit" (Vogel 2002: 33). The seizure of nature, which means the humanisation of nature, is a form of reproduction of reality. Until now, we could only bring to mind advances like cloning or transgenic food. But after the announcement that synthetic organisms have been created in the laboratory, there can be few doubts (moral judgements aside) that the human species has developed an exceptional ability to intervene in *deep* natural processes that we had deemed out of reach. Humans cannot create *ex novo* natural processes, of course, but they can manipulate and recombine them in a degree that is literally equivalent to the social construction of nature.

Deep natural processes? Indeed. Kate Soper's distinction between a *deep* nature and a *shallow* one is very relevant in this context. Whereas *deep* nature refers to the causal powers and structures that operate constantly in the physical world and are the condition for any human intervention in the environment or the biological realm, the *shallow* nature is that of our immediate experience in the everyday life: animals, the natural environment, our bodies, the material resources (Soper 1995: 156). A similar criterion is used by Dieter Birnbacher (2006) to tell nature in a *genetic* sense from nature in a *qualitative* sense. The former refers to the moment of nature's coming into existence without human intervention, whereas the latter alludes to the appearance of natural forms that can be, and actually are, affected by humans. Genetic nature's description is *historical*, qualitative nature's description *phenomenological*. Therefore, nature as an ahistoric essence is not the same as nature as a historic process. Whenever the autonomy of nature is invoked, this key distinction is lost, since deep nature's autonomy does not necessarily lead to the protection of superficial nature's forms. And even less so now that mankind is increasingly able to intervene on a deep natural level too.

Thus to take constructivism seriously means to recognise the role played by the ever-increasing interpenetration of society and nature, regardless of whether we *like* it or not. Since if mankind is natural, nature, due to the long process of co-evolution and reciprocal transformation which has been taking place since at least the apparition of *homo sapiens*, is also social. Moscovici has written about a "human history of nature" aimed to put an end to "the vision of a non-human nature and a non-natural man", insofar as nature and society do not exclude each other: "once society separates humanity from the animal realm and gives her a distinctive character, society becomes a second nature" (Moscovici 1975: 27). Only when we recognise that society has been originated in nature, but has separated from it *afterwards*, can we also speak of an *interaction* taking place between society and nature. Thus it is feasible to discriminate between the natural origin of society and its later historical development – along which society is constantly linked to its material base, but also gradually acquires a relative independence from it. And it is such separation that allows humanity to make sense of its place in the world, to adapt itself to it through the physical and symbolic appropriation of its environment. It does so despite the ontological unity between society and nature that underlies such *appearance of separatedness*.

Yet, despite the fact that society separates itself from nature through labour and culture, it is remarkable that nature and society come to be together again. However, this time it happens the other way around. The organic exchange between humans and their physical environment ends up transforming nature itself. Human environment is the final outcome of that process. Nature and society are reconciled again, in a sort of socionatural *end of history*. We are not back to square one, as an unrealistic green thought would expect, but the contrary takes place: if humans had belonged to the natural order, to the point of *being at unity* with it, now the natural order is absorbed in the social one, and the dichotomy society-nature – first a philosophical invention, then a social tool – is dissolved with the emergence of the *environment* as a synthesis. As humans depend on natural processes, natural processes depend on humans too (Eder 1996: 49). Nature and society are actively generated co-constructions (Irwin 2001: 173). The simple distinctions between "the social" and "the natural" are becoming unsustainable in the face of the complex and interactive character of social and environmental change (Redclift 1999a: 68). Not even the separation between the natural and the artificial, as we shall see, is easily set.

Human intervention in nature has been, and still is, so intense and massive that is does not seem reasonable to exclude this material dimension from any constructivist view. Besides the ontological reality of nature and the social character of our knowledge of it, we must take into account the crucial *practical* application of such knowledge *on* that very nature – and this application is also part of the social construction of nature. In other words: "Because the material properties of the constructed object still matter, so does the knowledge directed towards these material conditions" (Görg 2004: 30). More exactly, both a conscious re-construction of nature and a non-intended influence on it are to be found, the outcome of which is the social *assimilation* of nature and its transformation into *human* environment. David Kidner has referred to such a re-constructive task, but he has attributed it to an "industrialist fantasy" pointing out its limits in the face of nature's "ontological reality" (Kidner 2000: 346). Rather than capitalism, though, as he implies, it is evolution: humans appropriate themselves *naturally* of their environment and such appropriation (maybe fatally so) can only become more and more intense with the passing of time. For his part, Tim Rogers (2009) has defended a *relational* sort of constructivism that takes into account human-nature interaction. Be that as it may, green theorists will understandably refuse this conclusion:

> The interconnectedness of things is axiomatic in ecology, as in evolutionary theory, and environmentalists often appeal to it when making the case for human dependence on non-human nature. They generally have less to say about nature's dependence on us. But, of course, it works both ways (Jordan III 2005: 196).

As expected. After all, if the natural world becomes more and more human, if there is no nature left, but a mixture of nature and artifact, environmentalism could lose its purpose and hence its meaning. But is a constructivist approach immoral for that reason? The answer to this question depends on the role we assign to the social scientist – although it seems more immoral to ignore or distort reality in the name of, well, morality. In this regard, it is worthwhile to explore the kind of normative consequences that can be derived from the acceptance of a realist kind of constructivism.

Constructivism and Natural Limits

The problem of the natural limits to human activities is directly related to the clash between realism and constructivism. On the one hand, because this topic will be seen and weighed very differently depending on the approach, realistic or constructivist, that is finally chosen; on the other, since this position will result in potentially diverging politics of sustainability. It is precisely in relation to the latter that the shortcomings of both a realism without construction and a constructivism without reality are to be exposed.

Predictably enough, environmentalism has always defended an objectivistic approach to the design of human activity upon nature. Natural limits are obvious – the last piece of coal, the last well of water – and they point to weaknesses in the constructivist approach, once we leave aside the futile games of discourse theory. After all, it makes no sense to talk about social construction when the resources we need to survive have been exhausted. Or does it?

Such emphasis on the existence of natural limits to human activity has been a part of the green identity from the outset. In fact, a Malthusian core can be discerned in environmentalism, such is the faithfulness with which the core thesis of the British economist is generally endorsed. To Malthus, given that the growth of population is geometric and the growth of food resources are arithmetic, the available resources are a non-negotiable limit for social reproduction (Malthus 1999). But this is not the case and, as we pointed out earlier, it did not happen in his own time. His reasoning is, in fact, misleadingly simple and ultimately mistaken, since it does exclude the possibility that natural productivity can (humanly led) *improve*. Nature is not static, but dynamic, and human intervention in it intensifies such a trait. Marx's reply to Malthus points to this anti-essentialist stance. To him, there are no absolute *natural* limits, but rather *social* limits to human activity. These limits are historically-bound and temporary, so that nature's limits are relative rather than absolute ones. Thus the natural limits end up being social limits (Marx and Engels 2009).

This premise rang true during the neo-Malthusian revival provoked by the publication, in 1968, of a pioneering piece of green thinking: *The Population Bomb*. The author, Paul Ehrlich, alerted readers to the demographic trends of the time, which seemed unsustainable from the standpoint of alimentary resources. His book gave rise to a language of limits popularised a few years later by the famous report for the Club of Rome's on the limits to growth (Ehrlich 1969; Meadows and Meadows 1972). His warning did not lack foundation, but finally became what Stewart Brand called a "self-defeating prophecy" (see *The Economist* 2007). The reason is the successful green revolution fostered by Norman Borlaug's work and the subsequent increase in farming productivity. In other words, a social intervention that removed a (seemingly unchangeable) natural limit.

A realistic constructivism is thus better equipped to explain both the relative quality of natural limits and the human ability to go beyond them through the social re-construction of the natural environment. It is socionatural interactions that make such constant redefinition possible. That is why the last piece of coal will be probably substituted by a new energy resource. Whereas a realistic account is based on the idea that natural resources only admit one sustainable use, already settled in its biophysical features, a constructivist one states that there are many social ways of using the resources in a sustainable way (Acselrad 1999: 55). The very notion of *resource* has a marked constructivist emphasis, since it entails an instrumental apprehension of a given natural object that otherwise or the rest of the time is perceived in a different way (e.g. a tree) or simply neglected.

But does that mean that the natural environment impose no limit *at all* to human activity? Of course not. It is clear that the reciprocal co-evolution of society and nature can operate – regarding natural limits – both positively and negatively. Societies can waste their resources and disappear as a consequence. But recognising that nature sets some limits to humanity is compatible with the assertion that such limits are relative rather than absolute. If we insist on the latter, we end up talking about an unreal conception of both nature and society, seen as static entities whose interaction does not produce any changes in neither of them. Yet this is a view plainly refuted by history.

But it is not only society that changes – nature changes too. On the one hand, evolution takes place independently from mankind; on the other, though, social practices can produce unintended or unforeseen consequences, as well as *intended* ones, provoking further changes in the natural world. So:

> To say that scarcity resides in nature and that natural limits exist is to ignore how scarcity is socially produced and how "limits" are a social relation within nature (including human society) rather than some externally imposed necessity (Harvey 1996: 147).

Granted, human transformative powers are limited by the very *features* with which the environment itself provides – features that mankind can manipulate and recombine, but not elude. Human transformation of nature (its material and symbolic reproduction) does not take place in the void, but upon the pre-existent nature that constitutes his raw material. The outcome of such process will unavoidably *reflect* that very natural world. As John Stuart Mill puts it: "We can only take advantage for our purposes of the properties which we find" (Stuart Mill 1998: 7). A caveat that does not lessen human's enormous ability to transcend the limits that nature seemed to impose on them.

In any case, a realistic version of constructivism, grounded on the material dimension of the socionatural relation, helps us to understand – not necessarily rejoice – that both the natural limits that exist in a given moment in history and the social ways of dealing with the environment comes from the same place: the interaction between society and nature and the deep human intervention in the natural world. If we understand that natural limits are relative, we are at the same time acknowledging the dynamic character of socionatural interdependence, as well as emphasising the material, cultural and symbolic process it involves.

After all, the material conditions of social life change with time. Society transforms nature, but nature changes society as well. Only a realistic constructivism allows us to grasp that.

Constructivism and Intrinsic Value

The fact that constructivism may be a threat to the philosophical foundations of environmentalism is even more obvious if we put it in connection to nature's moral value. If there is a principle that has been fiercely defended by environmental philosophers, this is the principle that the non-human world has an independent value, i.e. a value that is independent from human judgement. A moral proposition follows, according to which nature, being an autonomous moral subject, must be protected for its own sake. Although environmentalism is remarkably diverse, the refusal of anthropocentrism and the recognition of nature's own value is a common feature for most of it.

The importance of such relation lies in the fact that any claim about nature's value must take into account the real conditions of nature's existence. Thus if nature does not exist in the very terms usually employed by environmentalism – as an independent entity, not created by humans, that humans cannot fully possess – it is not clear that we can maintain the invocation of its intrinsic value. In other words, nature's qualities can be tautologically described: "It's tangible, secure, rocklike, stable, self-evident, definable, real. In a word, it's natural" (Price 1998: 191). So: nature *is* independent, hence it *possesses* an intrinsic value. But if nature were a social construction, such value would be in doubt:

> Fundamental to this view is the assumption that there is a "correct" pattern of human-nature relations, independent of any human perspective. Otherwise this idea of transforming people's attitudes in the "right" direction would make no sense. But the idea of such an independent conception of human-nature relations is unintelligible (Levy 2004: 57).

Hence, if nature is not independent from humanity, does it still possess an intrinsic value? According to the latest theoretical explorations of the notion of the autonomy of nature, it does, insofar as it *exists*. As nature is self-sustaining and self-generating, its autonomy means that "what has come into existence, continues to exist, and finally, disintegrates/decays, thereby going out of existence, in principle, entirely independent of human volition or intentionality, of human control, manipulation, or intervention" (Lee 2005: 59). As David Heyd says, summarising the current status of reflection, recognising nature's autonomy means realising that it can maintain its organisation in the presence of external forces and that it may exert its own force on its environment while trying to maintain its integrity. It is a teleology that has nothing to do with mankind, but with nature's own goal-orientedness, so that "the issue revolves around the objectivity of values in the sense that something simply *has* a certain value just as it may have a certain other quality, such as a color" (Heyd 2005: 4–5). Ontology, in short, matters more than history.

Two fundamental questions arise. The first has to do with the kind of nature whose autonomy is thereby proclaimed. The second refers to the position wherefrom that very value is invoked. And neither can be properly straightened out leaving constructivism aside.

1. *Which nature is autonomous?* It is understandable that nature's autonomy ends up grounded on a teleological account of the former. Only an interpretation of nature that – despite its immanent appearance – is rather metaphysical can support the idea that nature's value lies in its autonomy from humanity. Because nature is *not* independent from us, in spite of the fact that it *was* once so. The nature whose autonomy can be recognised is meaningless for environmentalism, since it is the deep nature rather than the superficial one greens aim to protect. Certainly, the distinction between different natures is at work here: on the one hand, a deep/genetic nature whose (diminishing) autonomy can be recognised; on the other, a superficial/qualitative nature whose autonomy is refuted by the social history of nature. We cannot derive a particular duty of recognition for everyday nature from the abstract affirmation that it was originally created without humanity's intervention. As for its teleology, it is not clear why the circumstance that nature can sustain itself and leans spontaneously towards self-development should translate into the assumption of human duties of preservation. Nature has no plan, no teleology: it just exists. It can do so autonomously, if we do not touch it, but it cannot remain autonomous once we break into it. Thus the sense in which we can say that nature *is* autonomous becomes philosophically interesting, but practically trivial; whereas the sense in which we can say that nature is *not* autonomous is as philosophically interesting as it is practically relevant.

2. *Who says that nature is autonomous?* Not nature itself, certainly. However, environmental thinkers have usually maintained that nature's value is independent from human judgements. Nature does not only possess an *intrinsic* value, namely, a value that is not dependent on any other value, but also an *objective* value, insofar as it is a value independent from human moral evaluations (Siep 2004). Yet this is not, strictly speaking, an ethical statement, but a metaethical one: the statement that nature has a value which is "independent of the valuations of the valuers" (O'Neill 1992: 120). But this is not the case, since it is enough that one person claims the contrary, for such statement to be refuted. We do not even have human rights without an inter-subjective agreement on their creation and recognition! The most common objection to the intrinsic value of nature is implied, namely, that the very idea of value is a human convention, so that there is no such thing as an objective value. As Brennan and Lo (2010) have suggested, it is hard

to find any justification for the view that nature has an intrinsic value in the absence of a religious framework, especially as long as such claims are not supplemented by new factual (secular) information. A problematic trait of green realism – as opposed to constructivism – is involved here, i.e. a considerable deal of epistemological naivety, be it real or feigned. Actually, the role of language and culture in the human perception of nature is simply too important to be ignored. To come back to the comparison quoted above regarding colours, any quality of an object is humanly perceived, categorised, created: they are qualities of human's perception outside which the reality is meaningless. In order to distinguish colours, for instance, we need language and the representational mind language makes possible, so that we know what a colour is, as well as the distinctions *between* colours that make single colours distinguishable (see Sánchez Ferlosio 2009). There is a reality out there, a reality which is a condition of our existence that actually conditions us, but such reality is only properly experienced through the human categories that language brings into being.

Yet for environmentalism this leads to a *cul-de-sac*. More exactly, it leads to some sort of linguistic idealism which deprives us of any real foundation for the protection of nature. Reducing nature to a cultural construct is thus a direct threat to environmentalism's moral program, insofar as it suppresses the grounds for nature's intrinsic value. Philosophical arguments would have political and then ecological consequences.

However, being obvious that the most radical variations of constructivism are vulnerable to these arguments, such is not the case with the moderate version of it, and even less so with the realist version advocated here. The recognition that an independent nature (one that is *previous* to the social process of its appropriation) exists does not involve endorsing its independent moral value. On the contrary, insofar as natural and social histories converge, there is no autonomous nature. Once upon a time there was a more or less untouched nature, but it does not exist anymore, since human history has colonised it: a human history that is also, as such, natural history. Even those parts of the planet which are seemingly wild remain so thanks to human *omissions*. And that is so, it should be stressed, regardless of our moral stance towards it – regardless of how much we like or dislike it.

But does that mean that there are no reasons to defend the natural world or some given sustainable policy? Not at all. But we should be realistic about the character of that which we wish to protect. Nature is not autonomous anymore, but *we* are: thus we can decide that nature deserves protection, because we decide that it has some kind of value. Hence Neil Carter's suggestion that "a crucial defining feature of ecologism might be that it includes all perspectives which concede humans will always be the distributors of value, but they are not necessarily the only bearers of value" (Carter 2007: 36). The value of natural forms is decided upon *within* society. The divide ecocentrism/anthropocentrism would then be defused, making room for the coexistence of many different reasons for protecting the remaining nature – or not doing so, a possibility involved in the acceptance of "the ineliminable contingency of political argument" (Humphrey 2002: 2). In this regard, it will be suggested later in this work that, in view of the actual dissolution of nature into human environment, the reasons for protecting the remaining natural forms lie in their *difference* rather than in their *autonomy* before mankind. It is not quite the same.

Towards a Better Understanding of Socionatural Relations

So far it has been underlined that the frailties exhibited by the strong versions of realism and constructivism have given rise to some sort of pragmatic consensus around a moderate synthesis of them both: the object is real, the knowledge is constructed. However, it has also been pointed out that the resulting constructivism does not fully explain the material dimension of the social construction of nature – and that is why a qualified constructivism, based on socionatural interaction, is proposed. The former provides us with a realistic account of the latter. Such understanding is, in turn, a precondition for a feasible sustainability.

As it happens, the strongest versions of realism and constructivism stress only one of the parts of the socionatural relation – at the expense of the other. Nature is then either entirely natural or entirely social, but the fact of the mutual interaction of nature and society does not receive due attention. For its part, a moderate correction of these varieties of constructivism does take this interaction into account, but only in order to defend a somewhat *lessened* socionatural relationship, insofar as it does not give full force to the material dimension of it. Hence the pertinence of moving towards a realistic or qualified version of constructivism.

It should be stressed that the objective reality of the natural world is by no means questioned. For the social production of nature to take place *at all*, a natural reality must be there earlier. Therefore, the resulting re-construction will unavoidably reflect the character of those natural materials. In other words, the natural world does not *determine* the process of its social appropriation, but it does *condition* its development. We can only give form to the world we inhabit working upon that very world. Ted Benton is right when he cautions that, no matter how deep we go into the structure of those materials and beings upon which we work, human-made transformations *presuppose* and are *limited by* the powers and structures of deep nature (Benton 1992: 66). It is certainly the case that, although we are more and more able to alter nature's structures, we cannot *replace* them completely. However, it is not clear that this fact has the meaning that Benton suggests, since the existence of such ultimate limitation does not refute a well-proven and increasing human ability to transform (also deeply) the existing natural world.

Moreover, nature is not only a precondition for human labour, but also, immediately afterwards, a simultaneous outcome of their interaction: it changes together with humans and vice versa. In other words, if nature, as an objective reality, is a precondition for its social appropriation, it is no less true that the physical world *already transformed* is a part of the changing context of the ongoing social construction of nature. As Nikolai Bukharin puts it:

> The process of social production is an adaptation of human society to outer nature. But it is an active process. When an animal species adapts itself to nature, it is in fact submitted to the constant action of its environment. When human society adapts itself to its environment, adapts the latter to itself, so that it is not only an object of nature's action, but simultaneously and in turn transforms nature in human's labour object (Bukahrin 1972: 200).

It is an interaction that operates on several levels and directions. We transform a reality which, in turn, transforms us, but the product of that process changes the reality we work with – and so on, and so forth. Since such complex transformative action takes place on a pre-existent reality, maybe we should talk of a human re-construction of the natural world.

Furthermore, the process of human appropriation of nature is never closed. It contains two mutually bounded dimensions: the material and the cultural. There is a direct relation between the material processes (coal extraction, water desalination, DNA recombination) and the cultural ones (the study of the environment in order to unveil its laws, the attribution of meaning to those practices, their public or customary justification). The overall result of such relation determines the *shape* that the socionatural interaction adopts in a given time. As Michael Redclift says, we are materially and symbolically creative and destructive, we reorganise our environment cognitively and physically (Redclift 1999a: 67). Such reorganisation in turn influences society's culture, as well as its perception of reality and its attitude towards the future. They are mutually contingent processes, which shape themselves jointly, so that their *conceptual* separation does not hide the fact that their *practical* separation is simply not viable.

In this vein, we find with the late Karl Marx a concept that serves to portray socionatural relations brilliantly: that of human-nature metabolism (Marx 1993). It refers to the material exchange that, through the means of labour, constantly takes place between society and nature – a transhistoric phenomena, despite the fact that its particular expressions happen to be, well, within history. In a contemporary formulation: "Socioeconomic metabolism refers to flows of materials and energy between society and nature" (Fischer-Kowalski and Weisz 1999: 234). It is to be remembered that, according to Marx, history is to be conceived as the progressive humanisation of nature, which takes place together with the simultaneous naturalisation of humanity, meaning the realisation of humanity as a species. Socionatural metabolism is then "a social process of reproduction" (Bukharin 1972: 201). And indeed a dynamic one, full of tensions and imbalances between society and nature. So much for the greening of Marxism.

Yet a qualified constructivism entails an anti-reductionist stance. If the social construction of nature covers both the material and cultural sides of the wider socionatural process of co-evolution, it is not possible to give prominence to just one side of the process – be that the material or the cultural. To emphasise the socionatural interaction means recognising the complexity of a relation that cannot be reduced to "the material" (as Marx did) or to "the cultural" (as strong constructivists do). On the contrary, the material, cultural and symbolic features of that interaction are to be combined, since all of them serve to explain the wider process of social construction through which nature is humanised, until it becomes *our* environment. Since, as we shall see in the next chapter, human species has developed itself through its *differentiation* from nature, without ceasing to be a part of it, its adaptation to the environment has mainly consisted of the transformation of nature through labour. Hence we can talk of a social construction of nature, whose fulfilment is but the transformation of nature into human environment.

From this standpoint, society is a historical development *within* nature, one which paradoxically moves humanity *away* from nature. Yet such separation goes beyond a mere differentiation, insofar as it is at the same time an appropriation of the environment. We reach then another paradox: the human *separation* of nature is fulfilled through a more and more intense *interaction* with the natural world. Yet that may be only an apparent paradox. Is it not such separation a necessary precondition of socionatural relations? To say it with Niklas Luhmann (1989: 2), we face a binary opposition through which civilisation emerges as a counter-concept against nature. And the firmer such separation is instilled in us, the more intense the material interaction is and the larger the distance between society and nature. The frontier between them also becomes harder to settle, because there is no single way in which we, as humans, are related with the outer nature: the simple distinctions between the social and the natural are rather untenable (Redclift 1999a: 68).

Henceforth, the transformation of nature into the human enviroment puts an end to that historic process: nature and society get back together. But the differentiation between society and nature ends up reversing the starting point: in the beginning there was an undifferentiated nature, now we live in an undifferentiated environment. In this context, a conception of nature such as the one usually defended by environmentalism clashes with reality. Because only by removing the socio-historic dimension of nature can the natural world be portrayed as ahistoric and autonomous – an idealisation that can only be supported by a certain degree of sentimentalism, as expressed in the Arcadian views of nature or in the idea that wilderness is a refuge from the ugliness of civilisation.

Sustainability and Constructivism

The constructivist account of nature has been normally rejected in the green theoretical field. On the one hand, environmentalism reacts against the very suggestion that nature can be philosophically questioned; on the other, they fear that constructivism's spread can deprive environmentalism of its *raison d'être*. Thus the reply that constructivism suffers from a moral deficit as far as nature's defence is concerned, so that a sustainable society could not possibly be grounded on a constructivist account of nature.

However, whereas these objections may hit the mark as far as the strong version of constructivism is concerned, moderate constructivism is not so easy to refute – a fact that may explain the emergence of some sort of consensus around it. This consensus admits both the existence of nature and the mediated quality of our knowledge of it. Those who defend an Arcadian view of nature or the need for a human spiritual connection to it will dislike that *tertium genus*, but even for them is advantageous, since it preserves the useful delusion that an essentialist, ahistoric, universal natural world exists *outside* society.

Yet it is indeed a delusion. A moderate version of constructivism is not enough to fully explain socionatural relations in a realistic fashion. There are no such things as an outer nature and a social inner knowledge of it. Rather society and nature have been constantly interacting since the beginning of history, an interaction grounded on the process of human adaptation to the environment. This adaptation takes the shape of a human appropriation and humanisation of nature. Thus a realistic turn in environmental constructivism should enable us to overcome the dichotomy between object and knowledge, by incorporating into this picture the *material* dimension of socionatural relations. It is not only the concept of nature that is socially constructed: nature itself is constructed too. Society and nature have co-evolved by means of a *relational process* whose first consequence is the social appropriation of the environment. Human's transformative powers are greater by the day, hence nature and society grow more and more interdependent. Nature-society's dichotomy is *dissolved* into the environment. Stephen Vogel even says that *relation* is a misleading word, since "we are *in* the world" and "to *be* human is to be active in the world" (Vogel 2002: 32). Indeed.

Does this entail a moral deficit on the part of a realistic constructivism, insofar as it makes it seemingly harder to defend a sustainable society? Not at all. In fact, this question seems severely flawed. If realistic constructivism would say the *truth* about socionatural relations, or at least would make a proposal worth reading about them, should we ignore it, or silence it, just because their conclusions are incompatible with the arguments of those who defend *a* certain view of nature and *a* certain model of sustainability? In other words, should we sacrifice science to morality, pluralism to self-righteousness?

Besides, a material version of constructivism does not entail any obstacle to the achievement of sustainability. In fact, should it not rather help to achieve it? It makes no sense to defend a model of sustainability that does not reflect the reality of our exchanges with nature, as a qualified version of constructivism is able to do. There is no single sustainability, but several varieties of it. In order to choose between them – if such explicit *choice* is feasible at all – we must previously understand what we are talking about when we talk about socionatural relations. That reality be disagreeable is no reason enough to leave it aside! Sustainability is not about the protection of deep nature. It is about balancing society and its environment in a durable way. Protecting *and* restoring natural forms is not the same as protecting nature as a whole. In that regard, constructivism ends up being more realistic than so-called realism. Therefore, a realistic constructivism is arguably the only foundation upon which a post-natural sustainability can be built.

Chapter 2

Nature within Society

The Revolt against Dualism

Although it has just been shown that society lies within nature, having, so to speak, found its way into it via a long process of cultural and material appropriation, it can hardly be denied that nature is within society as well – since we are an animal species. This naturalness cannot be avoided, all the more since it is one of environmentalism's central tenets, if not the green claim par excellence: humanity belongs to nature and that entails moral duties towards it. In fact, the opposition between constructivism and realism can be seen as just another expression of the more general antagonism between mankind and nature, that is, of a philosophical dualism that separates the latter and thus plays a key role in shaping socionatural relations. In this light, such separation is untenable, since there are no scientific grounds to claim that mankind is different from the rest of nature – on the contrary, it comes from it, is a part of it, and depends on it. Thinking and acting otherwise would then be but a harmful delusion.

Needless to say, from its very beginning environmentalism has raised a powerful claim against any kind of dualism. From the green standpoint, the latter is an unfortunate cultural construction, a human convention that has served as a useful tool for nature's dominion and exploitation. The opposition between humanity and nature would have also provided a metaphysical justification for the subjugation of women and other human groups rendered as inferior (Plumwood 1993; Merchant 1996). Moreover, dualism can also be blamed for undermining the organic bonds between humans and nature, breaking up an original unity that actually leads to the death of nature and the subsequent human alienation from it. This is why the critique of dualism may be said to be environmentalism's first task.

This critique includes a tracking of cultural history that tries to uncover dualism's roots. Of course, it is all Descartes' fault. He has been demonised by green historiography as the father of the anti-environmental bias of modern philosophy and the bearer of an aggressive gaze to nature that has brought devastating implications (see Attfield 1994: 82; Oelschlager 1991: 88). Of course, there were early intimations: Plato's description of humans as heavenly creatures of immortal soul, Aristotle's distinction between form and matter, the biblical statement that humans reign over nature (see Glacken 1967). But the decisive moment for the advent of dualism is the Cartesian view of body and nature as mechanical realms opposed to the soul as the site for the rational (Descartes 2008). Although this differentiation was not new, Descartes organised the natural and human orders into a hierarchy: body *versus* soul (inside humans), humanity *versus* nature (outside them). This led to a *separation* of the human and the natural – a denial of human's naturalness. Later contributions, from Locke to Kant, just reinforced a separation that capitalism put into effect.

Arguably, the dividing line between humans and animals is the most representative form of dualism, insofar as it has traditionally served to regulate the socionatural relations by granting humans an exceptional status *above* nature. Such status has in turn come to define the very essence of the human. This sharp antagonism is also relevant on account of the peculiar position occupied by the animal itself, both socially and symbolically. Animals are living members of the natural environment, but also a part of human culture, be it as food or as companions, standing at the intersection of the cultural and biophysical environments (Dawson 1999: 193–4). In fact, the distinction between the human and the animal has also had important consequences *inter homines*, since any definition of the human that is based on the enjoyment of a particular quality turns those men who do not possess it into a sub-human or semi-animal being (Thomas 1984: 41). However, the intermediate cases of those feral or wild children that have not been submitted to any socialisation process have challenged the boundaries of the human and the natural throughout history, exposing how thin the line that separates mankind from nature actually is.

After Darwin (2008), this human exceptionalness became untenable. Natural evolution demonstrated the genealogical kinship of all living creatures, not just allowing a different interpretation of those bordering cases, but also demanding a thorough revision of the human-nature dividing line itself. Suddenly, although the recognition of this fact has been quite slow, mankind was not *better* than nature: it was just a part of it. As Mary Midgley puts it: "We are not just rather like animals; we *are* animals" (1995: xxxiii). This implies a different view of our relationship to fellow relatives. The fact that we are an animal species with a variety of interconnections with the non-human world, actually sharing a common descent with all other life-forms, is held to justify a number of moral responsibilities towards the latter (Baxter 2007: 1). Moreover, this paved the way for a naturalistic explanation of human beings, that is, an explanation based on a biological interpretation of human behaviour and culture. Any distinction between the natural and the social is to be abandoned, dualism becoming "the evil 'sin' to which any right-minded social thinker should object" (Newton 2007: 35). Society is thus re-absorbed by nature.

In short, naturalism is meant to overcome dualism. Whereas the latter sees a gap between mankind and nature that demands a separated approach to all human matters, irrespective of how that gap is explained, the latter claims that such a gap does not actually exist and stresses the need for a biologically-based analysis of human beings. The ensuing conflict among these approaches, as well as within them, revolves mainly around the degree of autonomy to be granted to human behaviour and thus to their collective expressions – culture and society. This conflict is far from being solved, despite calls to the contrary, exemplified by

E.O. Wilson's (1975) sociobiological programme, which paved the way for an explanation of human behaviour grounded on the same biological and evolutionary principles that explain animal behaviour. Later approaches, from evolutionary psychology to human behavioral ecology, have corrected the deterministic bias of Wilson's proposal, hence trying to explain human behaviour and cultural evolution through mixed models, wherein universal genetic structures are compatible with cultural processes non-reducible to a purely biological explanation (see Cavallisforza and Feldman 1981; Barkow et al. 1992; Pinker 2002). They all share the idea that there is a human nature, as opposed to the view that we humans are *boundlessly* malleable via environment and education – i.e. via nurture. As we shall see, however, we *are* malleable enough.

That said, environmentalism has not distinguished itself for advocating a naturalistic view along Darwinian lines. Although the wish to conjure social Darwinism may have played a part in such reluctance, the main reason for it lies surely elsewhere: the need to avoid any kind of behavioural determinism, lest the chance that human beings relate themselves with nature differently is frustrated. Environmentalism seems to have accepted just a part of the Darwinian legacy, the one concerning human beings, while sidelining the one that has to do with the rest of the natural world. Rather than Darwinism itself, it is the ecocentric interpretation of the latter that pervades environmentalism (Humphrey 2002: 24). The latter has proven to be more anti-anthropocentric than anti-transcendentalist. Above all, it has tried to contest the belief that human beings are exceptional. Yet this may be more than a simple belief.

Rethinking Human Exceptionalism

It is then true that we humans are not just rather like animals: we are animals. But it is not the whole truth because it does not explain how *different* we are from other animals and why that difference *sets us apart* from them, as well as from nature in a wider sense, thus turning an ontological identity into a historical separation. This is not to deny that we are biological beings, or rather, as we shall see, psycho-biological ones, nor does this mean a return to an untenable dualist position either. On the contrary, a naturalistic defence of human exceptionalism has been made possible by our current knowledge of the functioning of genes. Again, it does not follow from here that we cannot *assume* moral duties towards nature if we decide so, but it is far from obvious that we need to claim an identity with the rest of nature in order to proclaim such obligations.

It should be remembered that the Darwinist revolution does not only entail a demystification of mankind in relation to nature, but also, correspondingly, a demystification of nature and animals in relation to mankind (Janich 2010: 12). Neither humanity nor the natural world possess a transcendental meaning anymore. The romanticisation of nature is thus a paradoxical consequence of the scientific revolution, one, in fact, that environmentalism deepens via an ecocentric interpretation of Darwinism. But this romanticisation should not blind us: humans and animals are different in important regards. Moreover, an exploration of such dissimilarity enables us to see the history of the socionatural relation under a new light, which in turn may – quite unexpectedly – contribute to the refinement of such relation, removing human guilt and substituting it with a greater ecological awareness grounded on a nuanced affirmation of exceptionalism.

Now, for all the attempts to close the gap between humans and animals, the separation remains stubbornly firm. For instance, Donna Haraway (2007) has referred to our "companion species" and urged us to solve the philosophical questions they pose, while Jacques Derrida (2008) has claimed that the very concept of "animal" is a wrong conceptualisation of a living existence, blaming the Western thinkers for the fact that, unlike himself, "their gaze has never intersected with that of an animal directed at them" (Derrida 2008: 13). Yet this gaze is not so meaningful as our philosophical inquiries would like: this encounter is human-made, it belongs to our world of meanings, only we can formulate this interrogation. We can think about animals because we are successful animals. The idea that we can meaningfully communicate with our cat, as Derrida suggests, belongs to the realm of *poetry*. Needless to say, it is not a useless approach, but it does not shed as much light on human-animal relations as it is supposed to.

It was Martin Heidegger who famously opposed the poorness of the animal world to the richness of the human one. In his lessons about the fundamental concept of metaphysics (Heidegger 1995), he claims that the animal is poor in world, but humans are world-forming, whereas a stone is simply world-less. This statement does not derive from zoology, rather it is informed by it. Whereas humans can create their own world by extending the range of his relations with other beings and objects, animals lack both this range and depth of penetrability. This, Heidegger claims, is self-evident. Interestingly, though, he does not equate *richness* with *superiority*: "Every animal and every species of animal as such is just as perfect and complete as any other (...) this talk of poverty in world and world-formation must not be taken as a hierarchical evaluation" (Heidegger 1995: 194). Yet this proviso is intriguing. It is true that animal's poorness in world is defined *only* in relation to humans, so that in principle it does not involve any judgement about the nature of the relationship that the animal establishes with its environment (Elden 2006: 275). On the other hand, though, this comparison unavoidably entails a judgement regarding the degree of evolution and the range of possibilities that characterise humans *in opposition* to the animal. Otherwise there would be no poorness at all in the animal realm.

Heidegger was influenced in this regard by the work of the German biologist Jakob von Uexküll, to whom he just mentions *en passant* in his lectures. Von Uexküll (1970, 2010) stressed the importance of the *organism* for understanding any living being. Insofar as the existence of each species depends on different conditions, so Von Uexküll, each organism develops a different plan for dealing with the environment, so that each of them ends up actually inhabiting a different world. Their biological organisation differs, so their perception of reality differs as well. Thus each animal possesses a particular *Merkwelt*, or "perceptible world", constituted by the insights that *it* receives from the environment. Together with it there exists a *Wirkungswelt*, or "world of effects", comprising those outer objects adjusted to the tools with which the animal moves and feeds itself. Therefore, animals living in the same world of effects relate themselves to different perceptible worlds. In turn, what is perceived in each case depends on the complexity of the senses and of the central nervous system: most animals just see shadows, lines and colours, wherefrom some inputs are salient, especially those corresponding to food and threats. They are blind to the rest of the world. On his part, humans' greater biological development allows them to perceive a richer world. Above all, it is language that makes a crucial difference and creates an abyss between our perceptible world and that of the animal. Von Uexküll argues that our immediate vision gives us no chair, no car, nor any other object, but lines and colours which are turned into objects through *meaning*. We inhabit a different, richer world than that of the animal – yet we remain animals. Heidegger's insights are strikingly similar, as if they were a commentary *on* Von Uexküll:

> The lizard basks in the sun. At least this is how we describe what it is doing, although it is doubtful whether it really comports itself in the same way as we do when we lie out in the sun, i.e. whether the sun is accessible to it *as* sun, whether the lizard is capable of experiencing the rock *as* rock. (...) On the contrary, the lizard has its *own relation* to the rock, to the sun, and to a host of other things. One is tempted to suggest that what we identify as the rock and the sun are just lizard-things for the lizard, so to speak. (...) Every animal has a specific set of relationships to its sources of nourishment, its prey, its enemies, its sexual mates, and so on (Heidegger 1995: 197–8).

But let us leave aside this similarity. The point is that the difference between humans and animals is not just a matter of perception, but something that goes *beyond* perception itself, something that makes the former's perception qualitatively different than that of the latter. Having developed a complex language via evolutionary adaptation, an equally complex culture has emerged, wherein humans *live* and by which humans are *influenced*, not only in the course of a socialisation process, but each and every day of his existence. We are an *animal symbolicum*, as Ernst Cassirer put it (Cassirer 1972: 26). In the words of Terrence Deacon:

> The doorway into this virtual world was opened to us alone by the evolution of language, because language is not merely a mode of communication, it is also the outward expression of an unusual mode of thought –symbolic representation (Deacon 1997: 22).

This means that we are unable to experience events without assigning a meaning to them, attached as we are to cultural representations and meanings as determining factors of our behaviour and welfare (Castro Nogueria et al. 2008: 533). Certainly, this has to do with reason, but at the same time it goes beyond reason, to cover the language that conveys our feelings and affections, as Cassirer underlined. However, it is humans' unique ability to *step out* of a given course of living that allows them to develop a relatively autonomous culture full of meanings and symbols. Only humans are able to project themselves reflexively out of a given current situation, suspending his embeddedness, unlike the rest of animals, which cannot dislodge themselves from their biological embeddeness (Sánchez Ferlosio 1990: 214). They live, adopting Bataille's expression, "like water in water" (Bataille 1989: 19). For humans, this in turn entails a further ability, that of putting some distance between them and their bodies:

> On the one hand, man is irrevocably linked to his body, but on the other he is not identical with it, he differentiates himself from it. Whereas animals can be said to *be* their bodies, we must say that men *have* bodies (Bayertz 2005: 17).

Thus Peter Sloterdijk's conclusion: "We could even say that humans can be described as those creatures that fail in being animals, in remaining animals" (Sloterdijk 1999: 34). While not ceasing to be animals! Such failure also manifests itself in humans' ability to change the shape of their environment when adapting to it. They do so by creating an artificial, man-made world that sets them apart from nature – between the natural and the artificial realms. This self-development contributes to the richness of the human world, as opposed to the relative poverty of that of other animals. In short, humans are able to *create* their own world. Furthermore, as we shall see below, it is here that a human-nature dualism is originated: in the practical realm of socionatural relations. It is not, as Descartes suggested, an ontological separation – but an emerging, historical one. Henceforth, whereas the behaviour and development of particular animals can be described regardless of the times in which they live, those of humans will vary depending on their time, culture and contexts, and even on the individual features and abilities they happen to possess (Faber and Manstetten 2003: 170). A variability unknown in the natural realm.

Therefore, human reflexivity allows us to set our own goals, although these goals are not entirely independent from our biological configuration. They cannot be! Neither human beings nor their culture can be seen as separated from the former's genetic inheritance (Dickens 1992: 18). But they are not entirely dependent either. An enlightening debate follows from here regarding the significance of culture for the human–animal divide – or the need to dispose of it.

A good starting point is Ted Benton's (1992a, 1993a) attempt to develop a monist approach that is non-reductionist, i.e. a naturalist view that acknowledges the continuity and communality between humans and other species, while at the same time recognises human's specificity. He claims that although there are certain self-fulfilment needs that are peculiar to human beings, as self-conscious and historical subjects, they share several needs with other animals, *so that* the former should be seen as peculiar ways of doing what other animals *also* do. Therefore, needs that would seem to justify human exceptionalism should be contemplated instead as deriving from attributes that are common to humans and animals. Thereby, no human need can be seen as introducing a *qualitative* difference between them: insofar as both human and animal needs are biologically grounded, they are but different expressions of the same evolutionary process. This also means that there are not superior and inferior, but just different, needs. Thus the human being's distinctiveness is not to be found in the enjoyment of attributes that are lacking in other animals, but rather in the way in which the former satisfies needs that are actually shared with the latter. A difference in degree that *remains* a difference in degree.

But is it so simple? Some human features are not so easily reducible to purely biological grounds. In principle, as Benton suggests, the difficulty lies only with human needs that belong to the cultural realm, since physical needs (nourishment, reproduction) seem to be identical in both cases. Thus a distinction comes up between what may be called (i) *primary* needs, explained by referring to that what other animals *also* do, and (ii) *cultural* needs, in some sense consequences or manifestations of the latter. It is the relationship between them that is to be explained. Benton argues that the most sophisticated cultural needs must have appeared later in time, once the primary ones had been satisfied. But this does not seem to bring us very far. As Tim Hayward (1992) commented, if we accept that certain primary needs must be satisfied *before* cultural needs can be fulfilled, or even emerge, we are left with two alternative interpretations. Either fulfilling a cultural need is *in itself* the satisfaction of a more basic need (so that the former is in some sense a sublimation of the latter and the cultural need is not "superior" by any account), or cultural needs emerge as a consequence of the previous satisfaction of the basic needs, but they do so as *qualitatively* new ones (that is, autonomous and irreducible needs that cannot be explained as peculiar or indirect ways of fulfilling the primary needs). Whereas the first explanation leads to an exaggerated form of reductionism, the second seems too dualist for Benton and environmentalism writ large. What he appears to imply is that cultural needs retain elements of more basic needs, but they also add other elements that turns them into different needs, that cannot be reducible to the most basic ones, nor be explained without them.

Yet something is missing here. Human distinctiveness does not just lie in the emergence of cultural needs that are certainly non-reducible to basic ones, nor can be easily inferred from them. The difference that language and meaning bring about has also to do with the singular way in which humans fulfil *both* primary and basic needs. We could say that primary needs are humanly satisfied in ways that suggest a cultural and symbolic re-appropriation: sex is not the same as eroticism; we cook, not just eat raw food; and so on. The animal is there, to be sure; but the human is too. As Soper suggests,

what distinguishes the specifically human mode of gratification of needs held
in common with other creatures is the aesthetic and symbolic dimension itself,
and one must question whether a non-reductive naturalism of the kind defended
by Benton can fully respect this differentiation without falling into circularity
(Soper 1995: 164).

In other words, Benton is trying to explain the emergence of cultural needs as a
development of more basic ones, but in doing so he may be overlooking the fact
that the fulfilment of the latter is linked as well to specifically human features –
those, precisely, that other animals lack. Of course, language and consciousness
have traditionally sustained human exceptionalism. It could then be the case that
the so called logocentric Western tradition had indeed turned a difference in degree
into an absolute difference, in order to justify the domination of the natural realm.

This is why an elementary methodological caution should be noted. When
we make judgements about the inner life of animals or we attribute them certain
features or compare their enjoyment of some attributes with that of human beings,
we are doing so from *our* point of view using *our* cognitive resources (Perler und
Wild 2005: 13). Heidegger was very aware of this point as well:

> *Can we transpose ourselves into an animal at all?* (…) Thus once again we find
> ourselves immediately confronted by a *methodological* question, but one which
> is *quite unique in kind.* (…) Strictly speaking, therefore, this methodological
> question is a *substantive one* (Heidegger 1995: 201).

We cannot, at least as of yet, be certain about the accuracy of our statements. But
what does that mean? What are the implications of this limitation? Essentially,
it reminds us that our enquiries about the human-animal divide might be more
useful for the study of men than for the study of animals, since the latter are bound
to remain for a while (up to a point) *unattainable* to us. But again, that is not
so important in this context. The claim that humans have separated themselves
from nature while remaining a part of nature and that such separation is historical
and emergent, rather than ontological and absolute, is still valid. Furthermore, it
does not involve a definite judgement about animals' inner life or attributes, nor,
crucially, do they lead to moral rules on the treatment of animals. It is one thing
that we are different from them in some important regards; how we decide to relate
ourselves to them is a different matter. As Meyer suggests,

> no particular normative posture toward nature or toward particular types of natural
> beings or entities is a consequence of accepting a constitutive understanding of
> the role of non-human nature within human life and society (Meyer 1999: 18).

In fact, as an species, we have not had much of chance: struggling for survival and
welfare, we dominated nature in a way that now appears to be in some aspects –

among them, the treatment of animals – untenable, or, at least, inadvisable. But that is *our* concern. It could not possibly have been our ancestors'.

A nuanced naturalism can indeed help to make this point more salient, fostering some changes in this relationship *from now on*. That is: now that we are self-aware enough and possess the scientific knowledge to implement them without curtailing our welfare. Furthermore, although the fate of animals throughout history deserves our pity, we may come to realise that we humans can also be the recipients of such sympathy, strange as it may sounds. We were just immersed in the struggle to survive and thrive, we lacked the self-awareness that wealth brings about, as well as the moral and technological resources to be more compassionate. This feeling of pity towards ourselves may seem absurd from the point of view of an animal activist, for instance, but it might be valuable: it reconciles us with our past, a past of brutality towards animals and, alas, towards other human beings. It does so through *pity*, a sentiment that compels us in turn to be better – to animals and to ourselves. A species in the path to self-improvement and refinement: that is a narrative that may well resound in a society orientated to sustainability.

Despite this caution, we should not forget that this difference in degree is such that it *actually* separates human beings from animals and *actually* makes the former exceptional animals. In this regard, it comes as no surprise that naturalism's main difficulty lies in explaining human singularity without falling into some form of dualism, while evading biological determinism at the same time. But then again, that is not unfeasible. An evolutionary explanation of human behaviour and culture is compatible with the recognition of humanity's *exception* – that of a psycho-biological animal that is simultaneously inside nature *and* apart from it. This is a unique condition indeed, made possible by the relative indeterminacy and openness of human nature.

The Openness of Human Nature

How can we reconcile the view that human beings are "animals under influence", as Sloterdijk (1999: 17) puts it, with the apparently opposite view that we come to the world already conditioned by our genetic inheritance? If there is no such thing as a blank slate on which to imprint a freely chosen human socialisation, it appears that only a severe naturalism may explain human behaviour, forcing us to abandon our received wisdom concerning what means to be human. On the face of it, that should suit environmentalism, insofar as it seems to imply that we are indeed conditioned by our belonging to nature and would confirm the absurdity of anthropocentrism. Still, a deterministic view of human nature works both ways: it suggest that we are indeed tied to nature, but at the same time it entails that we cannot help but acting in a given way, which is, needless to say, perfectly natural.

Yet naturalism does not necessarily lead to determinism. Our current knowledge about the functioning of genes do not support the view that human behaviour and culture are strictly determined by biological inheritance. In other words, the fact that human nature consisting of a biogenetic structure actually exists does not mean that such human nature is completely impervious to the environment which it inhabits – including its cultural environment. Whereas the denial of human nature makes no sense anymore, it would be equally wrong to bet on a closed conception of the former. Because human nature *is* open. That is, not absolutely, just relatively open. But open enough for taking such indeterminacy as its most salient trait.

Dieter Birnbacher has convincingly explained how difficult it is to identify a human attribute that may serve as a marking feature of the species itself – apart from the very human ability to *change* itself. It is precisely this ability that makes any other attribute historically contingent. His words: "If man is a cultural being that stems from nature, then man's nature is partly, in a general sense, tantamount to his capacity to continually change his biological nature" (Biernbacher 2006: 180). Thus the open quality of the latter. Humans become entities whose identity is not simply given; rather they project themselves on an open future (Heyd 2005: 65). Therefore, humans' historical development is *in some sense* self-guided, something that for some reasons has came to worry Jürgen Habermas himself: "The uneasy phenomenon is the disappearance of the border between the nature that we *are* and the organic structure that we *give* to us" (Habermas 2005: 44). Of course, that this should be a concern precisely *now* has to do with the increase of our manipulative powers. However, a conservative reflex is also perceptible in the kind of moralising reactions that claim against an alleged humans' dehumanisation. But how is it to be decided what is truly human and what is not? In which moment of human evolution should we try to stop, forcing a quite unnatural steady-state wherein humans do not change anymore? This is most unattainable. Dehumanising can only mean a departure from a particular cultural ideal, it cannot carry a wider, universal meaning (see Passmore 2000: 444). Arguably, humans can try to influence their own evolution – although this influence has so far adopted the form of a spontaneous process rather than that of a self-conscious collective decision – but they surely cannot *stop* it.

Paradoxically, despite the inheritance of a certain number of genetic traits, the very process of humanisation can fail if individuals are not properly socialised, that is, if they are not subjected to the *influence* Sloterdijk refers to. The abovementioned case of the wild children superbly illustrates this point. In fact, their example does not only come to show the openness of human nature in a *negative* way, as failed human beings, but also in a *positive* way, since they do not cease to be successfully living beings – other than humans. Glossing the case of Víctor L'Aveyron, a wild child found in the woods of France in 1797, Sánchez Ferlosio (1990) has suggested that the situation in which these children are found should not be interpreted as just the result of a *lack* of human education, but also as the result of the positive *presence* of a different education and a different *acquired* condition, which actually constitute "a biological commitment as radical and profound as the human one" (1990: 189). The four paws of the wild child does not indicate a lack of learning, but a positive learning, the functioning of his abilities in order to survive in a given environment. Thus the native indeterminacy of human beings is in this case biologically determined outside the human community, the result being an other-than-human creature: an animal that is just like an animal.

Therefore, human nature cannot be denied, but neither can nurture. In fact, the dichotomy nature *versus* nurture that has come to summarise this whole biological-cum-philosophical quarrel is largely a false one. Genes do not code human features directly, but rather are a necessary, albeit not unique condition for the development of the latter (Baxter 2007: 183). Genes are bits of information that can be used in different places and times through the action of biological promoters. That is why although we share a great amount of genes with some other animals, the disposition and functioning of that similar amount of genes can produce a formidable difference on the kind of entity whose existence they organise. As Matt Ridley puts it, genes are designed to take their cues from nurture: "It is genes that allow the human mind to learn, to remember, to imitate, to imprint, to absorb culture and to express instincts. (…) They are active during life; they switch each other on and off; they respond to the environment" (Ridley 2004: 6). As of now, empirical evidence suggests a genuine interplay between genes and the environment (see Rutter 2006). There is thus a mutual influence, whereby genes condition but do not determine human culture, and human culture influences but does not determine genes. Culture can then be seen as a biological phenomenon made possible by the plasticity of human neurophysiology, the traces of which can be explored throughout history by distinguishing between life-regulating traits that are relatively stable and behavioural traits that admit a much greater variability and are influenced by culture and the environment (Smail 2008).

Therefore, it can be said that human culture is relatively autonomous from – or is relatively determined by – biological inheritance. But the contrary is also true: the tools developed by hominids provoked changes in their extremities, while the size of the brain augmented as well in order to accommodate the greater amount of information gradually stored in culture. The latter, in turn, becomes more and more complex through the intensification of human contact facilitated by communication technologies. It has recently been argued that the *exchange* of ideas and products between men is a major cause for the development of culture (Ridley 2010). This may be the case. After all, exchange can be considered an *accelerator* of cultural transmission, insofar as it connects previously unrelated groups, allowing them to gain access to successful ideas and goods developed elsewhere. Furthermore, both cultural transmission *within* groups and cultural influence *between* them introduce a *contingent* element in the evolutionary process, since the local success or failure of cultural variations is unpredictable (see Castro Nogueira et al. 2008: 26).

It can then be concluded that both human nature and human culture are relatively open. They are not completely determined by biological constraints and genetic inheritance, which however remain a major constraint on them. Nature and culture co-evolve. Moreover, a great deal of such co-evolution is propelled by the human effort to overcome the hardships imposed by the environment, that is, by the human appropriation of the latter, which in turn involves practical, linguistic and symbolic human actions. It is this adaptive process that in practice *produces* a dualism that lacks any ontological foundation.

The Emergence of Dualism

The recognition that the human being is a part of nature is thus not incompatible with the assertion of his uniqueness. He is undeniably a product of natural history and his most salient features, from language to self-awareness, can be explained as a result of his biological evolution. In this sense, all that is human is also natural. Yet, at the same time, a human culture has emerged that is incorporated into the evolutionary process. Without ceasing to belong to nature, humans become something else. Human uniqueness is therefore *embodied* in socionatural history.

However, although culture emerges in the course of the evolutionary process, it is not to be seen as a mere by-product of nature, nor can be just reduced to it. Culture is rather the human way of organising the relationship with nature (Harrison 1998: 426). This is an exclusive human trait: "The difference between history and the evolution of nature without humanity consists in the fact that no animal species is capable of taking charge of the objective conditions of its existence" (Godelier 1986: 67). A central feature of such organisation – which is always partly conscious and partly spontaneous – is the fact that the process of human adaptation to the environment includes the active transformation of the latter: an adaptation that entails a humanisation. The expression of the human species particular way of being involves the active transformation of the environment and the creation of its ecological niche (Barry 1999: 51). As Edmund Russell has underlined, social forces should be recognised as evolutionary forces, especially if we keep in mind that evolution is an ordinary process that happens at all biological levels on a daily basis (Russell 2011: 3). To be sure, this active transformation is not always conscious, but also unconscious as well as unintended.

Although we can underline the similarities between humans and the rest of nature, this hardly makes sense if we wish to explain the history of their mutual relations. Because it is human uniqueness that has made possible the domination and appropriation of the natural environment. Through adaptive domination, human beings set themselves apart from nature. Needless to say, they remain subject to the latter's laws, but are also able to change some natural conditions which would have seemed immutable – from contraception to genetic manipulation. The point is that this singularity does not require a retreat to an untenable philosophical dualism. It can be explained via a non-deterministic naturalism that recognises human singularity. This was, incidentally, the position adopted by Karl Marx himself, who suggested in his 1844 *Manuscripts* that by labouring upon nature humans are able to develope a life-species unlike that of other animals, insofar as they turn their productive activity into a means, not an end, of their existence (Marx 2009).

Be that as it may, the important point here is that dualism itself is an emergent order that is produced by human beings in the course of their adaptation to nature. There is no such thing as an *ontological* dualism, but there is definitely a *historic* dualism that emerges throughout this adaptive process. This separation is also accompanied by a *philosophical* dualism that rationalises such human departure and at the same time helps to achieve it by asserting human uniqueness. As Giorgio Agamben (2004: 21, 29) has argued, the very idea of human nature is grounded on that separation, the cultivation of which is assumed by the "anthropological machine of humanism" as a "fundamental meta– physico-political operation" for human's self-understanding. It is through such separation that humans can make sense of their existence, an existence, to recall Darwin's main point, that is primarily devoted to subsistence and reproduction: personal self-realisation on Saturdays came much later. Paradoxically, the fact that the naturalistic critique demonstrates that no dualist position is tenable does not abolish the practical separation that has become a *fact* – actually, the defining feature of socionatural relations. Yet, to deepen the paradox, the chance that this practical gap is overcome does not lie in a human *return* to nature, but in nature's *entry* into the social realm, the actual merging of the social and the natural, that is, the increasing hybridisation of the natural and the human.

Henceforth, it is the socionatural interaction that must be acknowledged as the key factor of this process. It is the *contact* between human beings and nature – a dynamic, transformative contact – that makes the difference. Whereas human beings are a part of nature, nature becomes a part of society as well. After all, only by acknowledging that society distances itself gradually from nature does an interaction *between* them make sense. Of course, the opposite can be argued. If human beings actively transform their environments, we should not refer to society's impact on ecosystems, as if they were two separate systems, because they are not (Harvey 1996: 186). In fact, dualism never existed; it is an invention, a cognitive tool that helps us to deal with our environment (Evernden 1992: 94). But although dualism is certainly a human *creation*, it is not any less real: we *are* separated from nature. It is not an ontological separation, rather an epistemological and practical one. But then again, that is not a small treat. To put it differently, being a product of history does not turn dualism into a mere discourse. The human-nature division has become real little by little, through processes such as the functional separation between the urban and the rural life, or the increasingly strong symbolic opposition between the rational productive activity and the natural world (Stephens 2000: 277).

What history shows is then an actual separation of society and nature, even if the growing colonisation of the latter by the former makes it increasingly difficult to distinguish between two realms that constantly overlap. Ironically, the historic deployment of dualism provides the conditions for overcoming it in the long term. As Raymond Williams (1980: 83) has noted, when social relations with nature become manifold, active, self-conscious and permanent, as they are now, the nature-human separation becomes truly, obviously problematic. Thus *philosophical* dualism is threatened. As it happens, though, the practical dissolution of dualism is far from reinstating the original order wherein humans belonged to nature. On the contrary, the dichotomy nature-society is resolved by the emergence of the *environment* as a category that comprises both nature and society. After all, we cannot easily distinguish between the social and the natural anymore: natural processes depend largely on human beings, as human beings depends on natural processes (Redclift 1999b: 68; Eder 1996: 49). Therefore, the abolition of dualism only makes sense in a situation where humans exert a conscious control of their interactions with nature.

But is the disappearance of dualism advisable? Environmentalists think so – hence their efforts to achieve it. But they might think twice. It is not just that binary oppositions of this sort are organising categories that shape our understanding and our inquiries about the world (Benton 1994: 29). Most importantly, any attempt to re-organise the socionatural relation is unavoidably grounded on a dualist perspective. The separation is a precondition of the damage as much as of the cure (Soper 1995: 40). Stressing the continuity between humanity and nature as a means to foster the latter's protection is not necessarily a good idea. After all, such continuity does not seem to grant a special or distinct value to the non-human world, whose value lies precisely in *not* being human. That is why otherness may be a better foundation than closeness for strengthening nature's preservation.

Nature, Dominion, Otherness

Provided that human history has been driven by an adaptive process to the environment that entails a growing separation from nature via the latter's domination, a self-conscious effort to protect the remaining natural world from extinction or damage may very well be considered a sort of evolutionary *refinement*. Humanity would act – she already acts – as nature's self-awareness. It is not exactly sustainability what is here implied, but the protection of natural forms and ecosystems. Obviously, a concern for the latter can increase the chance that the former gives due importance to such protection, instead of choosing another, less conservationist path. Yet, as we shall see, sustainability is not tantamount to nature's protection. It can strike a balance with natural systems without worrying too much about the preservation of landscapes or the protection of particular life-forms.

Yet nature *deserves* protection. This is simultaneously a trivial and a tricky statement. It is trivial, because almost anyone would endorse it without hesitation. But it is also tricky, since it does not possess a straightforward philosophical or moral foundation. Only recently have humans come to the idea that the natural world deserves any special treatment. Throughout history he has done just the opposite, with the exception of some animals and landscapes which happened to occupy a privileged place in particular cultures. In fact, this has not changed: everybody wishes to protect koalas but loathes rats. However, it can be argued that the current wish to protect nature is mostly related to the perception of its growing disappearance due to human colonisation. Actually, this is more evident with landscapes than with animals. A salient reason for protecting nature is hence the will to avoid its extinction. A simple, yet effective rationale.

A powerful, albeit eminently anthropocentric argument to do so concerns the symbolic loss that such disappearance would provoke: the symbol of that what is not human. As McKibben writes: "Nature's independence is its meaning; without it there is nothing but us" (1990: 58). Robert Goodin (1992) sees it likewise: nature provides a larger context within which we can place our lives. Needless to say, this does not amount to nature's disappearance, because, properly speaking, nature cannot disappear. Rather it is a particular idea of nature that vanishes, that of an entity which is *independent* of us. Interestingly, such idea of nature reinforces the cultural dualism that environmentalism is supposed to be against. If nature is independent, nature is other than humanity. Thus the urgent need to protect the remaining natural world.

Yet, on the other hand, neither this nor any other idea of nature can exist *after* nature. And we are actually living after nature. It would seem that in such a world nature can have no meaning, because a neutralised and subjugated nature is, properly speaking, meaningless. Or is it? There appears to be a relation between the human appropriation of nature and the suppression of nature's meaning. The less nature is a direct threat to us, the less inclined we are to believe there is a mystery in it. Likewise, the more we know about nature's functioning, the more we resort to a materialistic explanation of natural processes. However, whereas human's *opposition* to nature has played a key part in the formation of human subjectivity, nature's social *appropriation* does not necessarily deprive it of its symbolic potential. Very much on the contrary, it can be argued that there is a link between the dominion of nature and our ability to provide meaning to it. Rather than leading to a deprivation of meaning, dominion could be seen as allowing a more enlightened valuation of nature – that is, it could be regarded as the *precondition* for a more free understanding of the natural world. Could it not be the case that, once the constrictive force of nature is severely reduced, we can lead a less constrained exploration of its meaning? In fact, we can properly ask ourselves about the meaning of nature *once* the latter has ceased to be a heavy constriction to our survival and welfare. In Philip Sutton's words: "It seems that the less danger natural processes pose for people and the more their basic material needs are met, the more likely it is that they will take pleasure in nature" (Sutton 2004: 27). Distance makes a peaceful contact, as well as a quiet reflection, possible.

Still, in order to accept this standpoint, we should stop identifying nature with the wilderness, despite the special significance that certain sublime landscapes may possess. Nature is present everywhere: its meaning and symbolism are not confined to those landscapes or environments wherefrom humans are absent. Domestic nature is also nature; our bodies are nature; the weather is nature. In sum, nature is within society as well. After all, as we shall see, naturalness is a matter of degrees, since there are different degrees of human intervention in, and intermingling with, nature. Rather than the whole nature-society interaction, it is the type and character of particular interactions that gain significance, so that the distinction between the natural and the artificial is a continuum rather than an opposition (Cronon 1996c: 89; Stephens 2000: 271). Besides, those areas where naturalness appears in all its glory – i.e. the remaining wilderness – are not free of human control; rather such control manifests itself in them quite differently.

In this context, nature's protection is not well served by making humans closer to it, but rather by fostering a sense of wonder and distance towards the natural world. We are a part of nature, but we have separated ourselves from it: nature is other, therein lies its meaning. We would not like to live in a world where everything is human. A spontaneous reaction to this argument consists in complaining that it is difficult to feel nature as other now that it hardly exists anymore. This is partly true. But only partly, because there remains a good deal of nature out there for us to experience and to reflect upon if we wish. On the other hand, a melancholic ascertainment must follow: it is only after a certain degree of control has been gained over nature that we can start contemplating it in a different way. In other words, where nature remains a threat, a constraint to human development, there emerge a will to appropriate it. Hence the sentiment that nature is to be protected from disappearance could not possibly have emerged earlier.

Admittedly, this entails that otherness is to be understood differently now: it is not the feeling that a sublime nature should be feared, but the result of a *fearless* aesthetic and philosophical reflection. In this regard, aesthetics is actually more useful than morals. As Matthew Humphrey has observed:

> For all that environmental campaigners may invoke science and abstract ethical arguments in support of their claims, the politics of the environment are, inevitably, closely connected to how human beings perceive the natural world, and in particular how any part of that world is "viewed" (Humphrey 2008: 138).

Such is but another expression of the latter's dominion. Theodor Adorno once observed that natural beauty does not exist wherever nature imposes itself upon human beings or is intensely laboured by the latter (Adorno 1997). That is why environmentalism does not always endorse an appreciation of nature grounded on aesthetic grounds: it can be interpreted as anthropocentric and hence as instrumentalist (Cooper 1998; Lee 1995). There is indeed an opposition between the aesthetic and the intrinsic value of nature, the latter being independent from human judgement, as well as from any notion related to the beauty or ugliness of the natural world. However, such aestheticisation is unavoidable. It constitutes a precondition of the human consideration of nature, as much as it is made possible by a sufficient degree of control over it. And it is surely less self-deluding to foster a greater respect to the natural world on a realistic basis, making clear that nature is at the same time the place where we come from and a reality that remains to some degree alien to us.

Chapter 3
From Nature to Human Environment

The End of Nature

The end of nature? No matter how many times it has been proclaimed in recent years, the announcement that nature has ended still requires a careful elucidation. Such an emphatic categorisation is not to be dismissed on account of its seeming implausibility: the fact that the sky is still blue and the trees are left standing does not turn the end of nature into a simple philosophical provocation. It is not. Yet it is not a tragedy either, but rather the predictable evolution of socionatural relations. In other words, it is a fact we must come to terms with. We should not see this claim as a passing cliché or academic hype. But then a question follows: can environmentalism survive to the end of nature? If it turns out that nature has really ended, what do we need green politics for? The answer is sustainability – or more precisely, a post-natural sustainability.

But let us see. What do we intend to say when we say that nature does not exist anymore? Such a bombastic statement holds two different meanings which are closely intertwined. They are in turn limited by a third meaning in which nature is *not* abolished – a meaning that, for all its relevance, in the end remains paradoxically unimportant.

A first meaning concerns the *idea* of nature. According to this, nature is not what it used to be, because we perceive it differently. And the social understanding of the object modifies the object itself. We have neutralised nature as a threat, assimilated it as a resource, and isolated ourselves from it as a daily presence. This is a process of social appropriation which deprives nature of any independence – an independence upon which its meaning has rested so far. Bill McKibben said it early on: "Nature's independence is its meaning; without it there is nothing but us" (McKibben 1990: 58). Now that the social and the natural are indistinguishable, the latter cannot exert a normative function anymore: nature ceases to be a source of meaning. But it is not so much *the* idea of nature that disappears, rather *one* idea of it – one which environmentalism strives to preserve. Its core is the existence of something larger than us, a wider context for human's existence. Interestingly, the end of nature then comes to echo the death of God, since both are seen as external providers of meaning. In this regard, it is not hard to see environmentalism as "a religious quest", as Thomas Dunlap (2004) has suggested.

Still, it is noteworthy that this idea can also be welcomed, as Bruno Latour does when he celebrates the demise of nature as an all-encompassing concept that hides the multiplicity of beings and organise them into a hierarchy that makes human dominion feasible: "Thank God, nature is going to die. Yes, the great Pan is dead"

(Latour 2004: 25). From this standpoint, the end of nature leads to the emergence of many natures and the elegy is transformed into a celebration. However, it is not clear how this celebratory view can be reconciled with the second sense in which the end of nature can be interpreted, namely, a humanisation of nature.

Since the end of nature as an idea does not exhaust the meaning of this particular category, Ingolfur Blühdorn (2000: 40) is surely right when he says that the abolition of nature is not to be regarded as an event in the empirical environment, but rather as the collapse and replacement of certain patterns of *perception*. But he is right only on the condition that we see this replacement as the due adjustment of social awareness to social conditions: i.e. as the perception of an empirical change that has been taking place for centuries. Socionatural relations have changed, our cultural apprehension has not. Is it not precisely this cognitive adjustment which leads to the sudden invocation of the ecological crisis? It is indeed a late realisation that leads to cultural overreaction. Ursula Heise suggests that a common theme in all those authors supporting the claim that nature has ended is the realisation that Modernity's destruction of the latter "has reached a decisive turning or end point with the current ecological crisis" (Heise 2010: 16).

Henceforth, the second sense in which the end of nature can be stated refers not to the idea of nature but to nature's *reality*. Nature has morphed into human environment and it has done so via its humanisation. Nature was relatively independent from society; now, after history, they are interdependent. The limits between the natural and the social are blurred until they finally disappear. We are not short of concepts able to express this. It has been said that we live now in a "post-natural world" (McKibben 1990: 60), made of a "created environment" (Giddens 1991: 124), which has put an end to the antithesis between nature and society, so that nature is not understood anymore outside society and vice versa (Beck 1992: 80). Therefore, a trait that was exclusive to mankind, namely, the hybrid position between nature and artifact, now encompasses nature at large: somehow, we are all artifacts. Or hybrids. Needless to say, the interaction between nature and society has always existed. But it is the intensity of it that is unprecedented, as it is our control of socionatural interactions.

As a matter of fact, a new overarching concept embodying this very idea, i.e. that nature has ended and morphed into human environment, has gradually risen to prominence in the last decade: that of the Anthropocene. It was first proposed by Crutzen and Stoermer (2000) in order to capture the quantitative shift in the relationship between humans and the global environment, as provoked by the massive influence of the former in the natural systems that underpin the latter. This influence increases exponentially since the advent of the Industrial Revolution, which marks a moment of punctuation in a long history of human interventions in the environment. Thus the term Anthropocene suggests that the Earth is moving out of its current geological epoch (called the Holocene) and that human activity is largely responsible for this, so that humankind has become a global geological force in its own right (Steffen et al. 2011: 843). Global warming is the most spectacular outcome of this deep change, but it is far from being the only one.

Disappearance of pristine land, urbanization, industrial farming, transportation infrastructure, mining activities, loss of biodiversity, organism modification, technological leaps, and growing hybridization are also on the list. In this regard, the Anthropocene may said to constitute the geological translation of the idea that nature has ended. Interestingly, the concept is consistent with a refinedly Darwinian view of the human development on the planet, since it does not rule out the possibility that this whole process – which has arguably made a laboratory of the Earth (see McNeill 2000) – ends up being a huge human *maladaptation* with unforeseeable consequences.

Yet is this not anthropocentrical madness, as environmentalists like to point out? According to them, nature cannot be controlled nor abolished, since it enjoys a radical independence from mankind that no human intervention is able to suppress. Although the operations of nature can be influenced, this is far from affecting its existence – not quite the same thing. Besides, mankind has always transformed its environment, but has done so according to nature's laws. Thus nature, understood as a constant set of laws and powers, remains independent from human beings. There is no third meaning for the end of nature, because nature's *essence* remains untouched. Hence nature cannot possibly end. The metaphor remains a metaphor.

If only it were so easy! This latter objection is not as meaningful as it seems, since it is not able to counteract the novelty that the idea of nature's end brings about: we can discuss the ontology, but it is not the ontology that matters. On the contrary, what matters is the relationship between nature and society – as expressed in a multiplicity of particular relationships between nature and human beings. In this regard, the end of nature has a twofold meaning: (i) natural processes cannot be defined anymore as independent from human influence, except in the rather abstract sense of their ontological dimension; and (ii) natural forms and processes have been transformed and humanised to a very high degree. Again, the distinction between deep and phenomenological nature sheds light on the subject. We should respect nature's causal powers and structures,

> but we shall not "destroy" nature at this level if we fail to do so. Nature at this level is indifferent to our choices, will persist in the midst of environmental destruction, and will outlast the death of all planetary life (Soper 1995: 249–50).

Thus deep nature is the condition for something, instead of nothing at all, to exist; therefore, it cannot disappear, unless everything disappears. But what has it to do with the nature environmentalism wishes to protect? Not much. Furthermore, although deep nature is not threatened, its operations *can* be influenced by human transformative powers. So much so that we can hardly declare that it remains totally autonomous. This loss of autonomy is especially visible in the humanisation of the superficial nature. Sometimes, human intervention is manifest; sometimes, it is not. But it hardly makes a philosophical difference, as Stephen Meyer comes to lament: "What separates the Brazilian rainforest from New York's Adirondack

Forest Preserve from Manhattan's Central Park is only a matter of degrees" (Meyer 2006: 9). Human control means the end of nature as we know it.

Therein lies an important consequence for any politics of sustainability. Since deep nature is not at risk, what environmentalism tries to protect is *a certain state* of socionatural relations, a *moment* of socionatural history wherein natural systems are neither massively transformed nor fully humanised. A given set of natural forms should then be protected, namely, those manifestations of superficial nature that express an *ideal* of nature based upon its appearance of *naturalness*. After all, nature in a genetic or deep sense does not need any help from us. But, as it happens, the nature whose protection environmentalism is committed to *has* disappeared, *has* been assimilated by society, *has* been transformed into human environment. It is an uncomfortable truth, but one whose acceptance should serve for the renewal of green politics. Strangely enough, it is a renewal rather than a demise, because the end of nature leaves enough room for a politics of nature.

Nature against Artifact?

So far it has been suggested that nature does not exist anymore *because* it has been transformed into human environment. However, it has also been claimed that the natural world has not been completely humanised, since human control does not necessarily have to express itself on every occasion in the same way: we can choose not to intervene and hence not to humanise. Yet this leaves a pertinent question unanswered, namely, when does nature, once it has been touched by humanity, cease to be nature? In other words, it remains to be defined precisely what makes nature to be nature, and thus what makes it – for humans – valuable.

Nature is that what is not artificial. On the face of it, this seems like a reasonable starting point for isolating what naturalness consists of. It covers all those entities and processes that come into being or exist without any human intervention. On the contrary, what has been created by human agency is artificial. Henceforth, natural entities are not the result of human intentions; rather they exist independently from human designs or purposes (Katz 1993: 223–4). Robert Goodin comes to a similar conclusion when trying to set up a green theory of value, inasmuch as nature's value lies precisely in its naturalness, understood as humanity's absence: "What is crucial in making things valuable, on the green theory of value, is the fact that they have a history of having been created by natural processes rather than by artificial human ones" (Goodin 1992: 27). A plastic tree cannot have the same value as a real tree, since the latter has been created by human beings, whereas the former has gone through a historical process of creation independent from mankind.

Unfortunately, the apparent simplicity of this criterion is soon complicated by a great number of nuances that render the distinction almost useless. Why? Because, as we have already underlined, natural history is also a social history: a history made of countless socionatural interactions, full of particular human interventions, whose extent has grown dramatically in the last three centuries, spreading the human influence in so many ways – gross and subtle, visible and invisible – that it is now far from obvious whether humans are *absent* from a given natural process or a certain natural entity, or are not. Are pets natural? And other domesticated animals? What about the woods and rivers that are human-designed? Granted, they remain natural; but in a sense, they are not *that* natural. As Neil Roughley (2005: 144) cautions, if any human-made change is going to exclude naturalness, then there is almost no natural thing left on the planet. In this regard, Nicole Karafyllis (2003) has proposed the term *biofact* to name those entities whose origin and formation has been anthropogenically influenced, directly or indirectly, irrespective of the actual *visibility* of that influence. Human beings themselves are, of course, the first entity in that list, but plants and animals belong to it too.

It appears that a definition of naturalness that relies on the absence of every trace of human influence is not very helpful. Yet a plastic tree is not the same as a seemingly natural tree; likewise, the sand and water of a human-made beach are not necessarily as artificial as the beach itself; and so on, and so forth. Where is the line to be drawn? Maybe nowhere. It rather seems that nature, in this sense, is a *gradable* concept – insofar as it is a gradable *reality*. Thus the question about the naturalness of a given entity or habitat cannot be answered with *yes* or *no*, but rather with *more* or *less*, after its composition and history has been carefully studied (Birnbacher 2006: 6). The opposition between the natural and the artificial is then better conceived of as a continuum. An abyssal fish is at one extreme; a plastic fish is at the opposite end; a genetically modified fish occupies the middle. Different degrees of intervention come to express different kinds of socionatural interaction.

But is that the proof of a social failure in assimilating the natural and turning it into human environment? Actually, neither the social appropriation of nature nor its social construction involves an *absolute* humanisation of the natural. A thoroughly artificial world is as fantastic as a thoroughly natural one – although the latter existed once and the former may one day materialises! Be that as it may, the fact that some parts of the natural world had *apparently* remained less affected by the process of social appropriation, allowing us to distinguish between *degrees* of humanisation, does not invalidate the wider transformation of nature into human environment. Ironically, in fact, those parts of the natural world that have been less aggressively appropriated, or even left untouched by the hands of humans, are so thanks to a social decision, i.e. to a deliberate social omission. Such deliberateness comes to show that the cultural and symbolic aspects of that appropriation are being fulfilled, insofar as the conservation of natural forms still plays a part in the satisfaction of human needs – be they moral or aesthetic. However, these pieces of nature are not situated outside society, but within it: they are human environment too.

Towards a Post-natural Ecological Restoration

If nature can experience different degrees of human intervention, the question follows as to which kind or what extent of intervention is acceptable. Needless to say, acceptable in relation to nature's protection. Let us for now leave open the question of the *technical* substitutability of natural beings and processes, assuming for the sake of the argument that *every* piece of nature is potentially substitutable, to focus instead on the *moral* suitability of such intervention. If we locate nature's value on its naturalness, the question arises whether we can restore nature by artificial means. If restored nature is taken as a fake and hence deemed worthless, the margin of action for a sustainable society regarding the protection of nature seems dramatically restricted. So therefore a seemingly philosophical debate leads, once again, to policy.

Unsurprisingly, the more value we attach to pure nature, the more restricted the range of human interventions deemed legitimate will be, and vice versa. By pure nature we name here those natural forms whose history exhibits no trace of human intervention, excluding hybrid forms wherein that intervention is milder or hardly visible. The assignment of value runs parallel to the abovementioned continuum: a wholly natural tree is more valuable than a genetically modified tree, whereas a plastic tree lacks any value at all. It is noteworthy that the value of naturalness can be either ecocentrically or anthropocentrically founded. In the first case, the recognition of nature's intrinsic value demands respect for its autonomy, which in turn means that human intervention is not acceptable, lest such value is spoilt. In the second, the very existence of an untouched nature, a nature whose creation has nothing to do with us, provides us with a larger context of meaning. Either way, nature's naturalness is to be respected. But it is obvious that the second position makes room for at least some varieties of intervention and, crucially, for restoration techniques. Take Robert Goodin's position:

> If what we value about nature is that is allows us to see our lives in some larger context, then we need not demand that that nature be *literally* untouched by human hands. We need demand merely that it have been touched only *lightly* – or, if you prefer, *lovingly* – by them (Goodin 1992: 53).

Nevertheless, a deep green position must reject this invitation, because nature is not taken as an end in itself, rather as a means to human ends: it provides us, humans, with a larger symbolic context. It does not make sense to praise the naturalness of a natural entity if it is to serve human needs. From this point of view, nature's protection is not just related to the *present* superficial qualities of natural entities, but rather to its *history*, which must be a history free from human interventions (Elliot 1982, 1997; Katz 1985, 1997). An important part of nature's entities lies in their autonomous development, which is to be respected by *not touching* them. Yet this guiding principle – a moral duty to respect nature's naturalness – can be carried out differently depending on how severely we interpret that command of respect.

Nowhere is this more pertinent than in a particular kind of human intervention which can effortlessly be considered as the lighter, most loving conceivable touch: ecological restoration. That is, all those human actions orientated to the renewal and recuperation of damaged or degraded – or even disappeared – ecosystems, habitats, and natural species. Is ecological restoration commendable? Or does it just lead to a faked nature? If strict historical criteria prevail, restoration must be refused:

> The attempt to redesign, recreate, and restore natural areas and objects is a radical intervention in natural processes. (…) Once we dominate nature, once we restore and redesign nature for our own purposes, then we have destroyed nature –we have created an artifactual reality, in a sense, a false reality, which merely provides us the pleasant illusory appearance of the natural environment (Katz 1997: 105).

Even more bluntly, the idea that nature can be restored is "the big lie" (Katz 1992: 269–70). If the integrity of nature admits no human intervention whatsoever, restoration becomes humanisation. This will be especially the case whenever restoration tries to serve human needs, as the managerial approach to restoration allegedly does, inasmuch as nature is perceived not as an autonomous being, rather as an assembly of ecological functions (Bavington 2005). A strict approach, in sum, sees nature restored as a fake or an instrument – but not as nature anymore.

A milder approach starts from the recognition that different types of interventions can be distinguished, not all of which involve aggression towards nature. It may be said that naturalness is respected if human intervention does not entail a supreme determining or guiding influence on natural entities or processes (Hettinger 2005). Thus it is possible to conceive an ecological restoration that is benevolent and does not entail human domination (Light 2005). In this context, ecological restoration can be seen as a tool for returning as much wildness to natural processes as possible (Jordan III 2003). After all, if ecological restoration is refused and humanisation follows its course, the net outcome is clearly detrimental to nature's protection. As Basl has recently claimed: "It is perfectly consistent that an object be created by humans and that it promotes nonhuman values and even a world view that is non-anthropocentric" (Basl 2010: 146). Henceforth, this is a conditional, pragmatic and a non-principled concession: a lesser evil.

Nonetheless, this approach to restoration can also be judged too timid. It is not clear why human interventions orientated to restore natural entities, processes or habitats should have any limit at all. The goal is, *if we decide to*, not so much protecting naturalness, understood as the absence of any human intervention, but restoring an appearance of naturalness through artificial means. When animal suffering is not involved, there is no objective reason to consider that changing a river's course is morally worse than genetically altering a flower – it is hard to see a substantive difference between them. It goes without saying that a fake must be well accomplished to perform its function as a replacement of the original, so that not just any restoration will do. But as Dieter Birnbacher has claimed, the important thing is that naturalness can be *experienced* as such on a phenomenological level, irrespective of the historical record of the natural entity or landscape in question: "Nature's protection is then a compromise between the inherently conservative tendency of the idea of nature's protection and the chance of a more progressive development, enrichment and fulfilment of the natural environment" (Birnbacher 2006: 78). That is to say, a restoration instead of (or apart from) a conservation policy. And more precisely, the kind of nature's protection that fits in a sustainable society for the post-natural age.

Allocating Value

According to a green conception of naturalness, nature possesses a greater value than artifacts. Such value stems from the fact that no artificial intervention has been necessary for the corresponding natural entity, process or habitat to exist. Nature is, in comparison to artifacts, the place of value. Moreover, that is so irrespective of the foundation – ecocentric or anthropocentric – upon which such valuation is grounded: a plastic tree is usually deemed less valuable than a natural tree, whether the latter is valued for its own sake or for the service it renders to human beings. But should it necessarily be so?

It is not so clear why we should automatically ascribe a greater value to naturalness than to the products of human ingenuity. We have already alluded to Robert Goodin's position that the natural world provides us with a larger context for understanding our lives, avoiding the feeling that it is only *us* that exist. He adds:

> Nor do I mean to suggest by this that human creations in general have no value, or even that humanly produced replicas of natural products are without value. The green position is merely that those replicas have *necessarily less* value than the ones arising out of natural processes (Goodin 1992: 27).

He then goes on to compare natural entities with art replicas: as the original painting has a greater value than the replica, so natural entities are more valuable than their human-made fakes. Yet we should not restrict the comparison to human-made replacements of natural entities or landscapes. It is more enlightening to widen the scope, comparing the natural products of biological history with the artificial products of human history. In this, naturalness is systematically assigned more value than artifacts. There exists a "natural bonus" that systematically provides a greater value to natural *existing* entities before human *made* things (Birnbacher 2006: 22). It is obvious that this feeling underlies the entire green movement.

However, it is unclear why a greater value should be awarded to what has been created through biological, chemical or geological processes in which humans have *not* intervened, instead of assigning it to all that humans create. After all, we have been able to create *something* where *nothing* existed. Furthermore, why should we not praise the human ability to introduce freedom and unpredictability in a natural world that operates out of necessity? A person who needs glasses to see a hurricane should not value the hurricane over the glasses: a *human bonus* seems more deserving in this case. As it also is in many others. John Stuart Mill, reflecting the classical view of the Enlightenment, put it like this: "If the artificial is not better than the natural, to what end are all the arts of life? To dig, to plough, to build, to wear clothes, are direct infringements of the injunctions to follow nature" (Stuart Mill 1998: 20). It is the correction of nature that makes human life possible and richer.

But the artificial we take for granted, whereas the natural looks somewhat different. Nevertheless, this usual response depends on a hidden precondition: a sufficient dominion of nature that makes it possible the moral and aesthetic contemplation of it. Survival comes first! Stuart Mill suggested that too: "It is only in a highly artificialised condition of human nature that the notion grew up, or, I believe, ever could have grown up, that goodness was natural ..." (Stuart Mill 1998: 46). As it happens, it is the artificial, applied to nature, that brings about both survival and the possibility of an enjoyment of nature – or better still: the enjoyment of some natural forms plus the abstract appreciation of nature as a whole. Let us not forget that the ideal of nature can blind us to its reality: a virus is as natural as a bear. It simply makes no sense that we put a greater emphasis in a domain against which we have had to rise in order to make human life possible, instead of realising the formidable value of the artifacts we have been able to create. This reaction can only be considered an emotional response, derived from a sentimental ideal of nature based on *forgetfulness*: of the human dominion through artificial means that makes this very discussion feasible.

However, a relocation of value on behalf of human artifacts does not necessarily lead to a neglect of nature's protection. Firstly, because this protection can be socially decided upon *despite* the recognition that human artifacts possess a greater value, insofar as they allow us to survive and thrive and finally – perhaps – protect nature. Secondly, as we have already underlined, because it is precisely through artificial means – whether they entail a direct intervention or just careful research – that nature can be protected if we so decide and in the measure that we agree upon. There is no independent nature, save for the remaining deep nature. Hence the assumption that we can easily separate the natural from the human is not very useful:

> an assumption which is far more realistic than that of a pure and socially unreconstructed nature is one that assumes that the casual powers of organic and inorganic nature have always been incorporated into and modified by human social relations and processes (Dickens 1992: 83).

Henceforth, mutual interaction and interrelation is the starting point for the protecting side of any sustainable policy. Only in doing so can we distinguish between different degrees of humanisation, choosing to restrain human intervention in those habitats and entities whose appearance is more natural than others – i.e. they seem to be the outcome of an ecological history free of human influences. An active restoration policy should actively supplement that protection.

However, it is to be remembered that the actual implementation of both conservation and restoration policies, as well as their extension, are subordinated to the social agreement around them, be it an explicit consensus on their desiderability or an implicit acceptance of them. In this regard, as we shall discuss later, liberal democracy is ambiguous. On the one hand, it represents a limit to the idea that nature has a special value, since that will be so *only* if enough people believe it. Nature's value is not an absolute truth and cannot be imposed on others. After all, we are not referring here to ecological functions, which could well be excluded from the moral debate in order to ensure human survival, but to the protection and the enhancement of natural *forms*. On the other hand, though, the complete humanisation of nature can be legally hindered if we take environmentalists' right to realise their conception of the good as a limit to such disappearance, as Mathew Humphrey (2002) has suggested. Be that as it may, a relocation of value according to which human creations are considered more valuable than nature and as the means to nature's future protection seems like the foundation for a more realistic axiology.

Rethinking Human Domination

The critique of human domination over nature is surely one of the most recurrent topics to be found in the green literature. Domination is usually presented as the very root of the current ecological crisis, hence also the reason for the human alienation from nature. History is depicted as the stage for a war waged against the latter, whose predictable outcome is a socionatural relationship based on a violent subjugation rather than on a harmonious co-existence. This is both self-detrimental and morally wrong. And it is not a plausible foundation for a sustainable society, which will only be achieved by replacing domination with gentleness. All in all, a familiar rationale. But is it accurate? Maybe domination can be seen differently – maybe domination is a necessary condition for sustainability. Admittedly, not just any kind of domination, but a refined one: domination understood as the conscious control of our interactions with the environment. Let us see whether this is a plausible alternative.

It is hardly striking that this theme is full of Frankfurtian resonances. The relationship between nature and society is central to the negative triumphalism defended by Adorno, Horkheimer, and Marcuse. It is in the rather obscure *Dialectic of Enlightenment* (Adorno and Horkheimer 1997) wherein the argument of nature's domination adopts the theoretical shape that environmentalism will inherit. According to their authors, although the purpose of the Enlightenment is to free mankind from necessity and superstition through knowledge, knowledge morphs into power and nature is in turn reduced to an object of dominion. This cannot come as a surprise, since knowledge would *in itself* be a subjugating force. Therefore, the noble goal of human emancipation ends up in the disenchantment and reification of nature. In addition to it, the domination of outer nature is reproduced within society, so that some men are dominated by others. It is a double alienation: an alienation from nature and a social alienation. In sum, Modernity's failure stems from a failed relationship with nature. Tragically so.

This argument had to catch environmentalism's attention. To greens, domination is but the culmination of the old separation between humanity and nature, which places the former above the latter – a separation that can be traced throughout philosophy's history in the writings of Plato, Descartes and the like (Leiss 1974; Plumwood 1993). But now, the extent of human domination is such that it could end up being equal to nature's annihilation. As Kirkpatrick Sale recently put it: "It is this extraordinary dominance by one single bipedal species that has brought us to the present imperilment of the earth (...) we are headed toward *ecocide*" (Sale 2006: 3). There is thus a subtle link between the mistreatment of a donkey and the threat of global warming. And the link is the *ethos* of human domination over nature.

Yet an argument's reiteration over time is no evidence for plausibility. Domination has become received wisdom, an *idée reçue* that, despite its dubious plausibility, exerts a powerful influence on the green programme. This should change if a realistic and reasonable sustainable society is to be achieved. Again, such a reformulation depends on a more sophisticated understanding of the socionatural relations. In fact, the thesis of the human domination of nature illustrates superbly the green confusion between myth and reality concerning those relations. The reason is that human domination is depicted as a historic *contingency*, i.e. something that has happened, but might not have happened at all. In other words, human domination is a particular historical development among many others, so that socionatural relations could have adopted a different shape, for instance the shape of a harmonious co-existence between human and natural communities. It could all have been different, had mankind chosen *not* to dominate nature.

This is precisely the first step in rethinking domination: to dismiss the idea that it is a mere historical contingency, something that could have not happened. By contrast, human dominion of nature is rooted in the human *adaptation* to the physical environment: its material and symbolic appropriation is made of necessity rather than choice. For humans to acquire some relative autonomy from natural constrictions, dominion is a necessary precondition. It is *afterwards* that we can *choose* the extent and shape of such dominion – as we are doing *now*. Therefore, the human dominion of nature is a part of the wider human adaptation to environmental conditions, wherein an antagonistic relationship is established, that does not exclude cooperative or even symbiotic dimensions. Human emancipation, in turn, has less to do with dominion than with the particular *use* of that dominion. Thus the former is not exhausted by the latter, notwithstanding the fact that dominion makes emancipation feasible. The difference between the two is what we call *culture*.

Henceforth, far from being a human choice, domination is a transhistorical necessity, an inherent feature of socionatural relations. If we talk about socionatural history, we refer to the human penetration into the environment through nature's transformation, consumption and use. Although cultures can differ in their impulse towards domination and in their feeling of kinship to nature, history has always involved human domination of nature in various degrees (Sheldrake 1990: 26). As Maurice Godelier puts it,

> *human beings have a history because they transform nature*. It is indeed this capacity which defines them as human. Of all the forces which set them in movement and prompt them to invent new forms of society, the most profound is their ability to transform their relations *with* nature by transforming *nature itself* (Godelier 1986: 1).

Thus inhabiting the planet means humanising it through labour. A non-dominated nature existed only *before* humanity itself existed, since "once we begin to speak of men mixing their labour with the earth, we are in a whole world of new relations between man and nature, and to separate natural history from social history becomes extremely problematic" (Williams 1980: 76). Still, that does not mean that any form of domination is sustainable, nor that a public debate about it is out of order. On the contrary. But on the condition that such dominion is not deemed as an historical *accident* that should be repaired.

Nevertheless, environmentalists like to stress that human domination is simply a form of anthropocentrical delirium, since it is not within the scope of human powers to complete such a task. In short: "Nature's control is a dream, a delusion, a hallucination" (Katz 1992: 267). For once, there are natural processes which remain inaccessible to us. A circumstance which has been crudely exposed by global warming, an environmental problem created by the side-effects of domination, which cannot be expected to be dealt with just by increasing domination. If human transformative powers are limited, domination too.

Yet this reasoning is fallacious. Why must domination be *absolute* to be domination? There can be domination even though it is not complete and unlimited. The human ability to handle natural conditions may be *enough* to exert an effective control over them. It suffices that it is what Reiner Grundmann (1991a: 111, 1991b: 2) has called a *conscious control* of nature. Thus the fact that an absolute dominion is out of reach becomes irrelevant if enough dominion is exerted. Nature's processes and entities do not have to be thoroughly manipulated for that dominion to be carried out. Likewise, a domination so conceived does not have to be equated with nature's *destruction*, inasmuch as it can designate its active and conscious *transformation*. In fact, the history of socionatural relations is also the story of human stewardship and human-nature symbiosis (see Radkau 2000: II).

Is this not a form of domination? After all, the latter refers to an uninterrupted process of dialectic interaction between society and nature.

Thus seen, domination is but the control – a transformative control – of the human interaction with nature. Insofar as a conscious and deliberate purpose is applied onto an inherently dynamic relationship, domination acquires a *reflexive* condition that makes full sense in the context of a refined socionatural relationship. A heterodox Frankfurtian thinker, Walter Benjamin, saw it this way: "So technology, too, is not about controlling nature: controlling the relationship between nature and humanity" (Benjamin 2009: 114). At the same time, this dominion must be based on nature's own processes and powers: only working with the tools of nature can mankind exert control over the natural world. As Alfred Schmidt observed, any control of nature presupposes an understanding of natural relations and processes, an understanding that springs from the world's practical transformation (Schmidt 1971: 95). This is a sort of *domination by mimesis* which even the adversaries of the latter acknowledge: "We can learn from nature not only in a moral, but also in a technological way" (Ott and Döring 2004: 198). Baconianly obvious? Yes, but crucial anyway. Bacon himself: "Now the empire of man over things is founded on the arts and sciences alone, for nature is only to be commanded by obeying her" (Bacon 2010: 106). This means that the separation between humanity and nature does not operate practically; on the contrary, their relationship is more and more intricate – although maybe less intimate. In short, the humanisation of nature entails also the naturalisation of humanity.

Sustainability and Domination

Seeing domination in terms of a reflexive control of socionatural relations has several advantages. First and foremost, it is a realistic view, as it is based on a non-idealised account of socionatural history. Secondly, it is also a pragmatic approach, as it does not expect to abolish human domination by cleaning out the human from the natural. On the contrary, it acknowledges the need for a dominion which is enough to make human emancipation feasible, but it also believes that domination can be refined in the context of a sustainable society – a refinement, however, which will likely involve a *greater* human control of nature through research and technology. Finally, it can be grounded on a tradition of Western thought that also happens to be the right foundation for the kind of sustainability advocated here: the humanist one.

To be more precise, this tradition does not seek either pure dominion or pure disembodiment from nature (not even, as deep greens suggest, in order to regain nature by spiritual means). According to this position, nature is bettered and perfected by humans, as a means to their own betterment. Although it is certainly an anthropocentric view, it does not exclude the cultivation of ecological values as a part – or, we could say, a late effect – of such betterment. This rationale belongs to the Enlightenment and acquires afterwards a strong Hegelian-cum-Marxian accent. Instead of judging nature as harmonious or even as a teacher of mankind, it sees nature as that what humans have ceased to be when they gain autonomy from their needs and instincts. But although nature is to be completed by humans, for them to feel at ease living within it, nature itself is not to be seen in a Cartesian way, but in a Hegelian one, namely, as a dynamic body which is dynamically laboured by mankind (Passmore 1974: 212). Again, Stuart Mill is very clear when dealing with the subject, as he argues that nature – outer nature, but also human beings' own nature – is not to be followed but to be amended:

> ... the duty of man is to co-operate with the beneficent powers, not by imitating but by perpetually striving to amend the course of nature – and bringing that part of it over which we can exercise control, more nearly into conformity with a high standard of justice and goodness (Stuart Mill 1998: 65).

This is surely anthropocentrical arrogance in the eyes of environmentalism. How can nature possibly be *incomplete*? Nature is as it is, and as such must be left alone. Yet nature cannot be left alone, unless we disappear from Earth. Still, dominion being an unavoidable trait of socionatural relations, it makes sense to pursue a *better* dominion – a dominion wherein human ends and natural protection can be made compatible. Tim Hayward (1992) suggests that the humanistic approach to dominion is not necessarily anti-ecological, as far as it is not defined in promethean terms. To be sure, inasmuch as humans have to neutralise natural constrictions in order to fulfil themselves, a bit of prometheanismus seems hard to avoid. We are mixed up with nature, we both belong to the same space, we co-evolve with it. But now, thanks to the reflexive turn in the socionatural relations, that dominion can be made more subtle, less constraining, almost a work of art.

Let us be clear: Franciscanism is not the way out. It is through technological development that our relationship with nature can become healthier and cleaner, as well as, of course, sustainable. Synthetic biology, the digitisation of natural systems through green information technology, genetic engineering, smart systems of information: it is all about science, it is all going to happen. As Stewart Brand puts it: "Human nature doesn't change much; science does, and the change accrues, altering the world irreversibly" (Brand 2009: 216). Even regarding global warming there are interesting developments in the realm of geo-engineering and scientific weather modification that point to the aforementioned control of human-nature interactions: one that is reflexive and conscious enough. Although many greens would prefer it to be so, modern society is not going to be dismantled. But it can, it should, exert a more refined dominion. It will.

Yet it is plain to see that a refined dominion of nature is but another name for sustainability. Again: sustainability after the end of nature. Admittedly, under the influence of green utopism, we had grown accustomed to the idea that the sustainable society of the future would resemble a pastoral landscape of socionatural harmony, achieved through a de-escalation and divesting process at the end of which production, trade and even travelling would be dramatically reduced (see Kassmann 1997; De Geus 1999). Beyond individual beliefs and lifestyle, however, and for all the abstract beauty this ideal might contain, it is more a self-delusive dream than a feasible goal. A global sustainability for a complex global society must be highly technical, as well as rich rather than poor: otherwise it will struggle to be effective *and* consented.

Should we not thus conceive sustainability as the last step in the domination of nature? Is the reflexive organisation of socionatural relations not equal to the supreme form of control over the natural world? It is the very logic of sustainability – a conscious re-arrangement of the environmental conditions – that leads to this conclusion. Even Adorno acknowledged this: "The more purely is nature conserved and transplanted by civilisation, the more inexorably it is dominated" (Adorno 2001: §74). Needless to say, not all kinds of domination are sustainable; but any sustainable society exerts dominion over nature. Nevertheless, this is a dominion not conceived as the destruction of the latter, but as the human control of the socionatural interaction. In this frame, an ecological crisis is not perceived as the outcome of domination, but as the expression of a *lack* of it (Grundmann 1991a: 109). This can be seen as a sign that sustainability is a spoilt paradigm, co-opted and absorbed by the capitalist system (Blühdorn and Welsh 2007: 189). Or it may be recognised as a fundamental trait of sustainability itself, irrespective of the final shape that a sustainable society may eventually adopt.

Because sustainability *is* domination. It cannot possibly be grounded on a harmonious nature that is separate from mankind. It is, by definition, a *human* task. It presupposes contact, interaction, interdependence. Sustainability can only be achieved controlling the outcome of socionatural relations. Such a control requires knowledge, such knowledge involves dominion. Not exactly the old dominion, but a new, better one – and the more sophisticated, the better. In sum, sustainability is a legitimate project for the enlightened domination of nature.

Nature, Society, Environment: A Recapitulation

The foundations have been laid for a conceptual reformulation that helps us to clearly differentiate between nature and the environment. Although both terms are used as synonymous in the public debate about the ecological crisis, it is advisable to separate them properly, in order to rigorously discuss the kind of sustainability that we would like to achieve. Nature and the environment are not the same, because the environment is, in fact, what nature has become. The "objective" nature that existed long ago has been re-interpreted by one of its elements: human society (Pepper 1993b: 442). In other words, the environment is the product of the social history of nature, of its integration into human history, of its transformation through labour. By that material intervention, nature is humanised and the natural becomes social. This view can be partly found, for instance, in Martin Drenthen's account of landscape, the latter being a space covered with remains of past human-nature dialogues through which "culture and nature merged more or less organically into a meaningful whole" (Drenthen 2009: 290). Therefore, nature is incorporated into the environment.

Be that as it may, it does not follow from here that everything that is natural has become an artifact. Firstly, because from a certain standpoint everything remains *natural* anyway. But secondly, and most importantly, because there are varying degrees of humanisation: social appropriation of nature is not always complete. There are spaces, territories, ecosystems, that have been less humanised than others. Hence they keep a greater resemblance with the green ideal of nature. Yet their conservation is grounded on social decisions and serves social goals, no matter how we define them: environmental equilibrium, moral respect for some natural forms, the aesthetic or spiritual satisfaction of humans themselves. In fact, it has been suggested that only a controlled nature can be properly appreciated, insofar as it is neutralised as a threat and its difference from us may then be aesthetically – i.e. intellectually – grasped. Furthermore, this human appreciation is unlikely to emerge in a bountiful context, where the visible natural forms (to leave microbes, for instance, aside) abounds. Nature's scarcity reinforces its value. But again, that is not nature anymore, only human environment.

Henceforth, it is advisable to restrict the meaning of nature to those habitats or entities that have not been highly humanised and thus retain a greater appearance of naturalness than others. They are humanly controlled all the same, or will soon be, through research and knowledge. They constitute the remaining nature within the environment. A sustainability policy will choose whether this protection should be extended, so that natural conservation is accompanied by ecological restoration – i.e. a more active approach that blurs the distinction between the natural and the artificial.

However, the concept of the *environment* can still mislead us, if we do not make explicit its opposition to the counter-concept of nature. We talk of an environment because we cannot talk of nature anymore: the latter has disappeared, becoming the former. We cannot even *distinguish* between them without being incongruous, since the environment is what nature *was*. More precisely, the environment is nature *after* history. It is what nature has become through its historic relation to society. We should distinguish these two terms from each other properly if we are to submit ourselves to the reality principle regarding socionatural relations. After all, sustainability deals primarily with the environment, not with nature – although, of course, by looking at issues relating to the environment one becomes involved with nature too.

PART II
Sustainability After the
End of Nature

Chapter 4
The Principle of Sustainability

Sustainability Afterwards

Is sustainability too important an issue to be left to the hands of environmentalists? That may surely sound as an unfair question, since without environmentalism's cultural work sustainability would not be an issue in the first place. Yet being right – or partially right – for such a long time may have distorted the green willingness to change course when it becomes necessary. In this regard, it is the rise of climate change that has confirmed the uselessness of an otherworldly approach based on a multifarious goal-setting, including a wide protection of nature and the end of global inequalities, which has usually led to an utopian blueprint of the sustainable society. Suddenly, this does not work anymore. Sustainability must above all make sure that the human species can survive global warming: niceties can wait. A pragmatic turn in sustainability is taking place and environmentalism should be able to make a contribution to it.

However, the need for a more realistic approach to sustainability is not new. For too long has the concept been subjected to all kind of rigidities due to the sense of urgency that has permeated environmentalism since its inception. It makes sense. Where there lies the conviction that an ecological crisis is under way, no other sustainable society can be accepted but a pre-fixed, pre-ordained, pre-planned one. A certain state of socionatural relations is to be replaced by a different one, which is to be indefinitely maintained in the future. Furthermore, a feature of this sustainable society is a strong protection of natural forms, which in turn restricts the range of permissible human activities, pushing the model in the direction of a steady-state economy. The justification of such societal turn has oscillated between two kinds of self-imposed moral duties: the protection of nature for its own sake and the protection of nature for the sake of future generations.

Yet considering sustainability in this way produced an unexpected cultural side-effect, namely, it turned the sustainable society into a *myth*. The habit of representing sustainability as a final state of humanity, to be reached only when a set of pre-given conditions were met, created a gap between the current world of the citizen and the future world of the environmentalist: between the real and the utopian. All the more so since these conditions appeared to be unattainable (restoring an idyllic socionatural harmony) or undesirable (giving up travel, trade, technology). The objection that a change in people's preferences could be achieved through education only reinforced the utopian imprint of sustainability. Thus it started to resemble the Marxian classless society, a ghostly place that history – teleologically and escathologically conceived – should bring about sometime in the future. The distance between theory and practice widened and a wrong idea of sustainability became widespread, i.e. a strange theoretical abstraction having nothing to do with our everyday lives. To John Foster, this is a mirage that is embodied in the very logic of sustainability as we have understood it so far:

> the goal itself recedes as our advance towards it pulls back, so that we will have less and less prospect of ever overtaking it. This truly is the politics of never getting there (Foster 2008: 66).

Admittedly, a utopian blueprint is supposed to be useful. It should educate the public about the need for an alternative, and the exaggeration it contains might help to achieve more modest proposals: asking for everything as a means to get something. It is not clear, however, whether the radical vision has helped to bring sustainability closer. On the contrary, it seems to have deepened a theoretical misunderstanding about its character, namely, the idea that sustainable society represents a *rupture* with the one in which we live. Sustainability is thus depicted as a breakup, as an uncertain path into the unknown. Yet it is not –for it *cannot* be.

Sustainability is rather a process of reflexive re-organisation of the socionatural relation which possesses no pre-arranged shape. This will gradually emerge from the inner workings of the current society. After all, it cannot be conceived of *as though* we suffered a terminal ecological crisis, *as though* nature still existed, *as though* the socionatural hybridisation were not to have taken place. Or, for that matter, as though we would not live in democratic societies wherein changes have to be explained, consented, negotiated through. Therefore, sustainability has to be consistent with the conditions created by the social history of nature, that is, by the transformation of nature into human environment, as well as with our sociopolitical institutions. Otherwise it ends up being a cultural fantasy.

But is it not sustainability that becomes a fantasy if nature does not exist? That is the understandable green retort against any post-natural understanding of sustainability. Yet what happens is exactly the opposite: the end of nature *liberates* sustainability. Because the latter is not about protecting natural forms nor about making global justice, but first and foremost about making human society sustainable. A tautology? Not exactly. The suggestion that sustainability must simply be sustainable means that it lacks a pre-given content apart from assuring that society can support itself biophysically now and in the future. Since nature does not exist anymore, we are not bound to protect it. And if nature does not impose a limitation on the societal organisation, sustainability can adopt several shapes. Whereas it *may* protect the remaining nature, it may *as well* minimise that protection – and the same goes for social justice or whichever other value supplement that is added to it. Sustainability, in short, becomes a *framework* for the public conversation and action about the carrying out of the sustainable society itself.

Nevertheless, the consolidation of climate change as *the* environmental problem par excellence has disrupted our understanding of sustainability as of late. Sustainability is relocated into the wider context of the social response to global warming. Although it was supposed to contribute to the green classical concerns by reinforcing ecological awareness, climate change has operated very differently, since it has *simplified* both the justification and the implementation of sustainability. The vague urgencies of yesteryear have turned specific nowadays. Sustainability is about survival and survival is, well, about whatever works. In this regard, it is relatively unimportant whether global warming becomes as dangerous as it is depicted; what matters is that we are taking the threat seriously. Therefore, the need for a special justification of sustainability vanishes. Or more exactly: the general case for sustainability rests automatically upon survival, whereas specific versions of that general principle will still require a more elaborated justification.

Henceforth, sustainability *after* the end of nature and *after* climate change must be understood as a realistic principle lacking any given content, i.e. as a general orientation of society towards the re-organisation of its environment. This does not mean that sustainability lacks *any* content at all. Whichever shape adopts society in the way to sustainability, some minimum standards must be fulfilled. These standards are not peacefully set either: they must also be agreed upon. As we shall see, this debate revolves mainly around the substitutability of natural capital. But a difference should be made between two levels of discussion: the minimum conditions for sustainability and the general shape of the sustainable society that is to be built upon them. Sustainability is a general principle within which several sustainable societies are feasible. In fact, we can see the growing agreement on the need for sustainability that global warming is helping to forge as proof that, far from being located in a remote future, as classical environmentalism would have it, *sustainability is already happening*. A cultural turn is under way, after which the re-organisation of the socionatural relationship is incorporated as a major political goal.

Therefore, the fact that such a principle is adopted at all can be understood as a part of the general trend towards self-consciousness and refinement on the part of the human species. In this context, reframing environmentalism and renewing green politics means precisely adopting a post-natural stance with regard to sustainability, yet promoting it at the same time as a means for re-organising socionatural relations and restoring natural forms in an active, non-purist way. Instead of moving from nature to sustainability, we move from sustainability to nature – through the environment. Needless to say, this is already an specific conception of sustainability, to be defended before others in the public realm.

Sustainability as an Open Principle

The theoretisation of sustainability has been transiently affected by an apparent paradox: the more accepted the idea that societies had to be sustainable became, the more difficult it seemed to define the content of sustainability itself. Hence a decade ago it was customary to declare that sustainability remained a powerful idea but a vague concept, threatening to become a mere cliché, especially after it had become a fashionable term outside the green realm (Harrison 2000: 1; Drummond and Mardsen 1999: 2). Such were the bitter fruits of success for an idea that – no matter how deep in the past its roots may now be traced – was globally launched with the publication of the well-known Bruntland Report for the United Nations in 1987 (WCDE 1987; Grober 2010). Sustainability seemed to be at a standstill.

It was a false alarm. The situation has changed during the last decade, throughout which the debate around sustainability has grown more and more sophisticated. It has acquired a notable degree of conceptual richness with the result that the issues around which it revolves are now neatly identified – ranging from natural capital to substitutability, from ecosystem services to sustainability indexes. At the same time, new tools are emerging for implementing it that take advantage of the huge potential of information technologies. Sustainability studies are more focused and more attention is paid to practices, policies and measurements. It is thus understandable that environmentalism itself is increasingly centred around an encompassing concept from which *any* other green issue may – or should – be approached. As Tim O'Riordan puts it: "Environmentalism is morphing into sustainability" (O'Riordan 2009: 313). A different question altogether is whether sustainability is morphing into something that does not fit into the traditional green expectations about what a sustainable society ought to be.

Be that as it may, although we have amassed a great number of critical insights about the *prospects* for sustainability, this cognitive flourishing has *not* resulted in a unified theory of sustainability (Atkinson et al. 2007: 1). But that is hardly surprising. A unified theory of sustainability is almost a contradiction in terms, since the concept possesses an inescapable normative quality opposed to any reductionism. We can talk of *theories* of sustainability, but not of a *single* theory for it. Seen in this light, the aforementioned paradox disappears as such: sustainability is an intrinsically open concept, because it is better understood as a general principle than as a particular theory for the organisation of society. It is thus a principle *and* a framework.

But what is sustainability then? Let us say: any kind of socionatural relationship which is balanced enough to be maintained in the indefinite future. No more *and* no less. This is achieved by a conscious re-arrangement of the human environment through variable means. However, as such, sustainability is a neutral concept. It does not make explicit the way in which this equilibrium will be reached: it is a general orientation of society that requires a later specification. Sustainability is thus described as "the capacity for continuance more or less indefinitely into the future" (Ekins 2000: 70), or as "the capacity for continuance into the long-term future" (Porritt 2005: 21). It is not, as such, an objective concept:

> Sustainability cannot be determined objectively because defining sustainability involves value judgements with respect to which qualities of which resources should be sustained by which means, as well as for and by whom. Differences in human values make people's answers to these questions, and hence their definitions of sustainability, differ (Sikor and Norgaard 1999: 49).

After all, the societal adjustment to environmental conditions can be carried out in several forms. Hence a general orientation towards sustainability does not determine the critical features of each specific adjustment, i.e. the degree in which natural forms are going to be protected, the priority to be assigned to social justice, the political means through which sustainability is pursued. That remains open, uncertain, indefinite. Or, it remains open in the abstract level until social actors start to *practice* sustainability according to each own's view of it. Thus sustainability is a principle open to public discussion, not the planned imposition of environmental standards to the social activity. It is always necessary to respond to some key questions: what is to be sustained, for how long, for which reasons? Reasons are not unimportant, as we shall see: the fact that a given social arrangement may be *technically* sustainable does not automatically provide a *moral* duty to follow it (Beckermann 1994, 2002: 74). This means that sustainability should be seen as similar to concepts such as democracy, liberty, or social justice (Jacobs 1999; Bosselman 2008). A normative principle rather than a particular theory.

Therefore, sustainability is open to several interpretations, whose content depends in turn on value choices about the kind of socionatural relationship we wish to promote. Different values foster different sustainable societies – some of which are technically feasible, some of which are not. For instance, the total destruction of the planet's woods would arguably lead to unsustainability; and the same goes for a restoration of primitive conditions of life if we take into account the human price of the transition from here to there. In between, though, lie several possibilities. The social sciences cannot reach a consensus on the *right* model of sustainability, since the latter does not exist. There is a radical openness in the concept of sustainability: neither its content nor its form is settled in advance. Scientific knowledge does not suffice, notwithstanding its key role in providing environmental insights. Social practices and political decisions – loosely or heavily informed by scientific insights – count even more. If we do not recognise this, we may be misrepresenting the way in which any kind of sustainable society, bar an authoritarian one, actually takes shape.

It is simply that we cannot know which shape that will be. This limitation, which may also be a blessing, can be interpreted in two different ways. On the one hand, as a recognition that environmental issues confront us with "genuine and widespread *indeterminacy*" (Foster 2005b: 113). Hence we do not know what the result of our policies and practices will turn out to be. This is true.

A complete certainty does not belong to this world. But it does not follow from it that a cautiously restricted policy is the wisest or most effective means to achieve sustainability (Costanza 2000). Why should it? We can acquire *enough* certainty to promote sustainability, accepting that a complete possession of it is simply unattainable. If we let the open quality of sustainability be curtailed by a general precautionary *ethos*, we may never achieve it. Yet no call to caution can stop the practices of innovation that – for good or worse – are driving the search for sustainability nowadays. In this regard, on the other hand, sustainability is meant to *remain* indefinite, that is, open to change and constant adaptation. In Susan Baker's words:

> Sustainable development is a dynamic concept. It is not about society reaching an end state, nor is it about establishing static structures or about identifying fixed qualities of social, economic or political life (…). It is better to speak about *promoting*, not achieving, sustainable development. Promoting sustainable development is an on-going process, whose desirable characteristics change over time, across space and location and within different social, political, cultural and historical contexts (Baker 2006: 7–8).

A word about sustainable development and sustainability. Some authors distinguish between the two, either suggesting that the latter is a goal whereas the former is the process by which we move towards that goal (Porritt 2005: 21), or underlining that whereas sustainability belongs to the realm of ecology, sustainable development is a social and thus a broader concept (Baker 2006: 7; Moffat 2007: 319). It is also frequent to find comparisons between different types of sustainable development that are interchangeable with others where the term sustainability is used instead. So, are they the same thing? It depends. They can be considered synonymous, if both refer to the general orientation of society towards, well, sustainability. But if we describe sustainable development as a particular path to sustainability (featured by an emphasis on development rather than growth and an acute concern with social justice), then they cannot be synonymous: sustainable development would just be a particular conception of the general principle of sustainability.

Be that as it may, such features – indeterminacy, dynamism – reinforce the idea that sustainability is an inherently, radically open principle for guiding social action. It is a principle insofar as it signals a general orientation for society: being ecologically sustainable. But it is not a theory nor a policy, although its practical implementation consists of a discussion between different *theories* of sustainability, each representing different sustainable societies to be carried out through different social practices and policies. It is implied in this discussion that any proposed sustainability must comply with some minimum ecological standards – i.e. it must work. As we shall see, this discussion takes place both literally and indirectly: it includes communications with performative value as much as actions with communicative valency. It will increasingly be, if it is not already, the extension of the long-standing debate over the good life *and* the good society.

In fact, it may be not so much an extension of such debate rather than the framework for it. Given the far-reaching implications of any re-organisation of the socionatural relations, it makes sense to conceive sustainability not only as a realm wherein different conceptions of the good make different choices regarding the shape of a sustainable society, but as the very *content* around which that debate will mainly take place in the near future. Thus sustainability is both a *principle* guiding social action towards a balanced socionatural relation and a *framework* for the conflict between different variations of that general principle. The shape of a sustainable society will be the outcome of such conflict.

The Forms of Sustainability: Weak *versus* Strong

Sustainability is henceforth a general principle for whose achievement no single formula exists. Henceforth, the different variations of sustainability will be particular interpretations of a shared goal: a balanced socionatural relationship. These variations respond to different value judgements about the good society and result in different sustainable societies – or rather in different blueprints for them. The latter can be in turn more or less dissimilar among them, depending on how deep the differences are in their core values. They can as well be very marked in some regards, but less in others: social justice can be promoted in a society that is also interested in deploying a wide protection of nature as much as in one that is less interested in that protection. And so on, and so forth: there are many aspects of society wherein the differences among several ideal sustainable societies can be spelled out. However, it makes sense to pay special attention to the very dimension that explains the existence of sustainability in the first place, i.e. the socionatural one.

This is precisely what has happened in the sustainability debate. Although social justice seemed for a time a criterion as relevant as nature's treatment for characterising sustainability (see Dobson 1998), and despite the usefulness of multi-dimensional approaches (see Baker 2006; Davidson 2000), the socionatural dimension has ended up being the most important criteria for sorting out different types of sustainability. After all, any conceivable sustainable society is grounded on a given "socioecological regime", that is, on a specific pattern of interaction between human society and natural systems (see Fischer-Kowalski and Haberl 2007). Needless to say, paying preferential attention to the natural dimension of sustainability does not preclude the distillation of some other key features of any sustainable society – such as its moral foundation, the type of economic development that it favours or its political arrangements. Yet it seems to me that the most important decision to be made regarding sustainability concerns nature.

More exactly, what is to be decided upon is the degree in which nature is going to be protected. In other words: the degree in which it is going to be substituted by human-made capital. From here results a typology that roughly distinguishes between a *weak* and a *strong* sustainability, adding sometimes two more radical proposals, i.e. a *very* weak and a *very* strong sustainability. The crucial element is the substitutability of natural capital: a weak sustainability accepts *prima facie* the replacement of natural capital for human-made one, whereas a strong sustainability, on the contrary, restricts it as much as possible. It is noteworthy that the adjective itself underlines the socionatural criteria, since what is weak or strong here is the protection afforded to nature. This is because in the end, it is the view of the society-nature relation that determines the preference for a strong or a weak sustainability. As Hediger puts it:

> Differences among these concepts are a consequence of different visions of a sustainable world, differences in the understanding and representation of economic transformation processes (production functions) with different assumptions about the substitutability of man-made and natural capital, and differences in the way the environment and its functions are perceived and valued (Hediger 2009: 31).

Whether we wish to protect nature for its own sake, to maintain a given level of social welfare for the sake of future generations or just to make the economic growth climate-friendly enough to avoid a future catastrophe, the key decision remains *how* to maintain a stable pool of resources in the indefinite future. Do we preserve natural resources or do we replace them with human-made ones instead? To answer this question, we must go deeper into these concepts. But not without a caveat: for all their abstracting force, models can be deceptive. In the real world, we shall never find pure embodiments of neither the weak nor the strong sustainability. As John O'Neill cautions: "Weak sustainability is rarely as weak and strong sustainability is rarely as strong as each of its critics maintain" (O'Neill 2009: 286). Reality will be much more mixed and hybrid than its current representations.

The concept of natural capital was pioneered by some ecological economists in the late eighties (Costanza and Daly 1992; De Groot 1992) and since then has become a widespread tool in sustainability studies, given its usefulness for depicting the socioeconomic usage of nature and for bringing environmental issues into economic thinking and decision-making (see Ekins et al. 2003). To begin with a general definition, natural capital comprises the sum of natural resources that are employed by humans for human ends. This subservience is what turns nature into a resource in the first place. Such is the case even though nature does not even have to be touched to perform such function – as is the case with the satisfaction of moral or aesthetic human needs, which require precisely the opposite, i.e. nature not to be touched. Needless to say, there is no such thing as a directly observable natural *capital*: natural capital is a metaphor that tries to convey the usefulness of nature for human ends. Or else a meta-concept that isolates different functions and benefits provided by nature (Ott and Döring 2004: 176).

Nonetheless, this metaphorical quality can be troublesome, especially since the trope involved intends to grasp nature as a human resource. After all, this is an anthropocentrical mode of speech that clashes with the usual understanding of nature among environmentalists. If nature is described as a form of capital, we may be misrepresenting or downplaying some key ways in which we are related to it, thus reproducing the reductionist and utilitarian view of neo-classical economics (Foster 2005a: 28). In turn, it is suggested that this view may hinder nature's protection, insofar as it implicitly assumes that the latter can be substituted and reproduced by other forms of capital (Victor 1991). It is to avoid this bias that the need for including non-economic usages of nature in the concept of natural capital is emphasised (Chiesura and De Groot 2003). Still, it is hard to dismiss the idea that the term "capital" is hardly full of ecocentric connotations.

Natural capital is then a particular way of conceiving the vast reality of nature. However, for the concept to become operational different types of natural capital are to be distinguished – otherwise it remains too large a category to be useful. A rich theoretical reflection has provided insights enough on this issue to elaborate a plausible typology.

In that regard, let us first remember that this particular sort of capital is but a part of the total capital on which human welfare depends. Apart from natural capital, this includes both the opposite notions of *human-made* capital (comprising artifacts, inventions, as well as human/social capital itself) and the hybrid *cultivated* capital (including domesticated animals, cultivated plants, and the like) [Holland 1999]. Obviously, the making of human-made capital is *also* dependent on natural components: a book needs paper and the paper comes from trees. Cultivated capital itself is based on the human re-orientation of natural living beings. Such clumsy caveats point to an important limitation of the notion of natural capital when it comes to separate nature from society. But more on this later.

Within natural capital, the contribution of each part of nature to human welfare or even human survival is taken as the most relevant criteria for further distinctions. More exactly, it seems that we can distinguish between *disposable*, *fungible* and *critical* natural capital: the first is irrelevant, the second is not that important, the latter is irreplaceable. I say *disposable* because a part of the natural capital that is not critical may be replaced for human-made capital, but it may also be the case that we do need to make this substitution, so that such natural capital is simply exhausted and forgotten. This means, in short, that we should not equate *irreversibility* with *criticality*: although irreversible natural capital can be critical too, it may not necessarily be so. The loss of a single animal species is surely irreversible, and maybe the function it performs in a given ecosystem is not exactly replaceable either, but that does not make it necessarily critical – *we* can get along without it. In short, irreversibility does not guarantee criticality.

Henceforth, critical natural capital seems to comprise those natural beings and processes whose loss is both irreversible *and* irreplaceable in terms of their contribution to human existence and welfare. The environmental functions they provide cannot arguably be substituted by human-made capital. A problem immediately follows: "With the present uncertain state of knowledge about ecosystems, and environmental functions generally, it is very difficult to judge which are critical and which are not" (Ekins et al. 2003: 173). Thus neither irreplaceability nor criticality are to be considered closed, absolute categories.

Drawing on previous insights by Georgescu-Roegen (1971) and Faber et al. (1997), the proponents of the so-called *Greifswalder approach* to strong sustainability have proposed a distinction between *stocks* and *funds* as a means of improving the typology of natural capital (Ott and Döring 2004; Döring 2009). The rationale is simple: while stocks are inevitably consumed when used, funds can be used without being consumed. Stocks include coal or oil and provide benefits in the form of *flows*. Funds can in turn be non-living (sun, air, water) or living (livestock, plants) and provide benefits in the form of *services* (Döring 2009: 132). This distinction should help to calculate irreversibility *and* irreplaceability in a more rigorous way, avoiding a homogenising conception of natural capital. Yet it might be necessary to add a more relational perspective, wherein ecosystems as such are also represented, since the web of relations they sustain is increasingly recognised as a delicate part of the socionatural balance.

Still, it is assumed that human survival or economic welfare cannot be the only criteria for protecting natural funds, since we must also be aware of the moral or aesthetic welfare of future generations of humans. Therefore: "This implies that the concept of critical stocks does not only apply to funds which are relevant for the survival of humans but is interlinked with the preservation of a broad variability of species" (Döring 2009: 134). This is a change of perspective that should lead us to underline the *natural* instead of the *capital* whenever the notion of natural capital is involved (Ott and Döring 2004: 202).

Furthermore, human welfare is not the only game in town. We have mentioned above that the notion of natural capital is bound to provoke discontent among environmentalists, insofar as it does not help to dispel the widespread assumption that nature is there for the benefit of human beings. On the contrary, environmentalism has always claimed that nature exists for its own sake. If we measure natural capital in regard to human survival and welfare, only human needs and interests are being taken into account. The natural world is just contemplated as a resource, in terms of its ecological functionality: "A problem in a great deal of discussion of sustainability, and environmental policy more generally, is one that defines the environment only in functional terms" (O'Neill 2009: 297). Yet functioning is not the same as existing. But is it possible to accommodate an ecocentric position within natural capital? Holland (1994) tries to do so, suggesting that an additional category of natural capital can be isolated, one that leaves functionality aside and recognises the intrinsic value of nature. He refers to those ecological units that are representative of each particular historical form of association and their historically specific components – a sort of living memory of natural evolution. He calls them "units of significance" (Holland 1994: 178). As Andrew Dobson (1998: 54) remarks, the vocabulary of substitutability does not apply here, since these "units of significance" are *by definition* unsubstitutable. A strong conception of sustainability, insofar as it is concerned with nature for nature's sake, must then include such units – or a similar notion – within natural capital. We will return to this later.

Interestingly, sorting "units of significance" out of the totality of nature implies recognising that although nature *in toto* may be worthy of protection, such protection is simply not viable. And the same goes for the distinction between disposable and critical natural capital. How are we going to tell one from the other? In other words, where does the critical quality of natural capital lie? To answer these questions, it seems unavoidable to focus on the goods and functions or services provided by natural systems – mainly but not exclusively in their relations of material exchange with social systems. Goods are a resource; a function or service is a structural condition for human life. Whereas the goods are provided by the ecosystem *components*, the functions are performed by the ecosystem *processes* (De Groot 1992). From here on, the current consensus depicts the following critical natural functions:

1. *Regulation functions*: regulation of essential ecological processes and life support systems (bio-geochemical cycling, climate regulation, water purification, etc.);
2. *Production functions*: harvesting from natural ecosystems of, for example, food, raw materials and genetic resources;
3. *Habitat functions*: provision by natural ecosystems of refuge and reproduction-habitat to wild plants and animals and thereby contribution to the conservation of biological and genetic diversity and evolutionary processes; and
4. *Information functions*: provision of many possibilities for recreation and aesthetic enjoyment, cultural and historical information, artistic and spiritual inspiration, education and scientific research.

As we can see, there is an obvious difference between the first three categories and the latter. The amenity services of natural capital are meant to embody the way in which nature satisfy important human needs that belong to the psychological or spiritual realm (Chiesura and De Groot 2003: 224). We will see below, however, that the critical quality of such functions – or at least of a number of them – is not so easy to prove.

The link between the catalogue of ecological functions and sustainability is not hard to grasp. A certain amount of natural capital must be preserved that guarantees the constant performing of all those functions. Nevertheless, this does not amount to the discovery of sustainability's philosopher's stone. Even assuming a general consensus on the recognition of such functions, the amount of protection to be granted in order to guarantee the right stock of natural capital remains uncertain. Or put differently: the natural beings and processes that constitute the critical natural capital are not automatically deduced from the identification of the functions that they provide. Moreover, the critical quality of both ecosystem components and processes does not in itself dictates anything about their substitutability.

Maybe we can someday reach a scientific consensus about critical natural capital which is empirically grounded. Indeed, the conflict between competing views on nature's replaceability would end if it could be determined which assets are critical (Atkinson et al. 2007: 4). Maybe. But that consensus might be unattainable, since the direction and reach of future science and technology is unpredictable, so that it is impossible to know in advance what will *become* substitutable with the passing of time – not to mention that the very possibility of such calculation is severely hindered if we take the so-called *information functions* into account. People may have different, even incompatible views on these latter functions.

Be that as it may, we do not possess enough certainty to identify *critical* natural capital nowadays. Not that this would close the conflict between different sustainabilities – it would just close the conflict about the minimum standards of protection that *any* variation on the general principle of sustainability should guarantee. In fact, the current conversation about critical natural capital blends scientific judgements, moral values, and ideological positions. The subsequent lack of certainty leaves room for different propositions about the amount of natural capital that deserves to be labelled critical to be advanced. As nobody would defend a model of sustainability that leads to human extinction, the central point in this debate ends up being the degree of substitutability of natural capital, i.e. the amount of natural capital that *may* be substituted as opposed to the amount of it that *must* be protected. The opposition between a *weak* and a *strong* model of sustainability comes to summarise – as well as to simplify – the choices at hand.

a. On the one hand, *weak* sustainability means a large rate of substitution and thus a greater restriction on the recognition of critical natural capital. It is intended to maintain a non-declining stock of *total* capital, irrespective of the way in which that goal is achieved. This is to say, it admits all kind of trade-offs between human-made and natural capital. Natural capital is more important as capital than as nature. Hence the largest possible substitutability is admitted in principle, excluding, of course, the critical natural capital as it is defined in each historical moment. Whether natural capital is at the end massively substituted or not will depend on the politics of sustainability, as we will elaborate later, i.e. on the ongoing process by which the conflict between sustainabilities is solved. But weak sustainability does *not* restrict in advance the scope of nature's replacement by human-made capital.

b. On the other hand, *strong* sustainability entails a restricted substitutability and a larger protection of natural capital. It is intended to maintain a non-declining stock of *natural* capital. Thus not all kinds of trade-offs between natural and human-made capital are acceptable. Nature is to be preserved for its own sake as well, or at least on account of a number of non-economic services it renders to humanity. Substitutability is accepted but severely limited. However, the precise amount of natural capital whose integrity should be preserved must also be decided upon. What is critical exactly? To protect the *irreversible* natural capital is certainly not the same than to extend the protection to nature's *units of significance*. The latter plausibly amounts to a *very* strong model of sustainability.

But which one is right? It could also be that neither is. But if this is the case, is there an alternative to this widespread theoretical opposition?

How Much Nature Do We Need?

The answers to the question of how much natural capital we need oscillate between an appearance of neutrality and a surplus of morality. As we have seen, it involves a concomitant decision on natural capital's degree of substitutability, i.e. on the amount of natural capital to be protected. What counts here, though, is not so much the natural *stock* as the ability of the natural stock as a whole to perform the environmental *functions* which are essential to human welfare (Ekins et al. 2003: 173). Hence enough natural stock must be protected which will keep those functions unaffected in the foreseeable future. But how much natural capital is that? And which one exactly? It is in this point that uncertainty meets morality, revealing that the radical openness of the sustainability principle cannot be so easily overcome.

Ecological economists are understandably cautious regarding a high degree of substitution. They underline the complementary relationship between natural capital and all kind of human economic activities. It is not only that producing human-made substitutes require inputs of natural capital, nor that economy itself is about transforming materials into good and services and consuming energy in the process (Cleveland, Hall and Kaufmann 1984; Costanza and Daly 1992). Above all, they point to the multi-functionality of natural capital. The latter provides key functions for human existence and well-being, ranging from the supply of resources to the neutralisation of wastes and the sustaining of ecosystems' health. Thus it is difficult to substitute natural with human-made capital. This multi-functional quality turns out to be the crucial non-moral argument against substitution, because it operates a change in perspective that makes substitution hard to defend. How so? We just said that the importance of natural stocks lie in their ability to provide services or perform a number of environmental functions instrumental to human welfare rather than in their individual value. But if such ability does not depend on given sets of natural stock either, but on the *structural relationship* between ecosystems, their substitution becomes too uncertain and hence too risky to guarantee the maintenance of those functions. Substitution becomes a dangerous affair and should in principle be severely restricted.

Thus substitutability has nothing to do with *pieces* of nature, rather it is all about natural *relations*. Or, if you like, about complex biological systems:

the deep rather than the shallow nature. A fine example of that is biodiversity. Far from just alluding to the degree of variation in living animal species and microorganisms, it is concerned with several ways of differentiating natural forms – where they come from, how they are related to each other, in which habitats do they live – and with ecosystems themselves. The functions that biodiversity perform are so manifold and complex, that it is even doubtful that the concept of natural capital can, or should, cover biodiversity as a whole (Ott and Döring 2004: 212). The concept of *resilience* has been put forward in this connection. It is a structural stability concept based on the idea that multiple locally stable ecosystem equilibriums can exist (Van der Bergh 2007: 67). It is a holistic notion: "The resilience of an ecological system relates to the functioning of the system, rather than the ability of its component populations, or even the ability to maintain a steady ecological state" (Adger 2007: 78). Sustainability should then be related to resilience. And it should take into account both the uncertainty attached to complex natural processes and the threat of large-scale and irreversible losses of natural capital (Dietz and Neumayer 2009: 269). The conclusion is neat: "Strong sustainability claims that the set of *natural* goods should not be reduced or diminished over time but overall kept constant" (Ott 2009: 54, my emphasis). Hence sustainability must be strong: a large substitution does not seem to be an option. What is implied in this set of arguments is that replaceability is not a matter of *choice*, but a dangerous option to be approached very cautiously. Not a normative, but a technical question.

But what if we *learn* how to substitute ecosystem's functions? This is hardly unthinkable. After all, it should not be necessary to replace complete ecosystems to achieve that end – it would suffice to make some amends on them. Yet that would undoubtedly damage their integrity. Even a more thorough substitution, or, more plausibly speaking, a more thorough hybridisation between social and natural systems may be reached. In that case, the degree of substitutability would actually become a matter of choice. Certainly not a complete replaceability, but a significant one.

It is here that the moral position against substitutability steps in. According to this, we should define nature in non-instrumental terms. Substitution would be then highly restricted, because we cannot substitute the natural goods upon which the fulfilment of crucial dimensions of human well-being depends. On the contrary, maintaining the capacity to appreciate the natural world and to care for other species requires the protection of particular environments. John O'Neill thus concludes:

> Sustainability, on this account, is a matter of protecting critical natural capital – that part of the natural environment which cannot be readily replaced – and a set of constant natural assets that do have possible substitutes through re-creation and translocation (O'Neill 2009: 297).

In fact, this is neither a technical nor simply a moral argument: it is an ontological one. O'Neill delves into the metaphysical arguments about naturalness and chooses natural goods' history as the source of their meaning and thus their value: "The history and processes of their creation matter, not just the physical attributes they display" (O'Neill 2009: 297). Moreover, they may not just matter for satisfying human needs, but also for protecting nature's intrinsic value, which is, needless to say, the ecocentric argument against substitutability. It is an argument whose validity does not depend on the technical ability to replace natural goods or systems: "Substitution *could* be made, but *should* not" (Ott and Döring 2004: 107). If only because nature cannot be replaced – it is only *lost*. Strong sustainability is thus the one which takes nature's protection into account, irrespective of future changes in our technological capabilities that may render current limits to substitutability meaningless. It chooses a moral rather than a cognitive ground as a criterion for deciding on substitutability.

However, properly speaking, sustainability is not conservation. It may include the latter, but it does not have to. A strong version of sustainability does not defend nature's protection on behalf of its ecological *functions*, but in order to preserve its *meaning* – a meaning that lies in a historically grounded integrity. It is relatively unimportant here (but not unimportant over all, as we shall see) whether such preservation is meant to fulfil human or natural interests. Now, sustainability is in principle foreign to the problem of nature's preservation: its goal is to keep environmental functions, not natural forms. In fact, a type of sustainability may be conceived wherein a protection of irreversible natural capital is accompanied by a general disregard for nature's units of significance, resulting in a general loss of nature's diversity. Yet that does not mean that sustainability must be indifferent to the natural world *qua* natural world. On the contrary. But that protection does not belong to the principle's core. There is no *necessary* link between the general principle of sustainability and the conservation of nature. It follows therefore that the more orientated to a principled protection of nature a version of sustainability is, the more acute the resulting normative conflict around it will be. Or even more precisely: the more openly normative such conflict will be. Because it is hard to see technical arguments as completely value-free.

In sum, there are three ways in which the defence of a strong version of sustainability becomes dependent on normative judgements about the most desirable socionatural relationship. Firstly, whenever the amenity functions of ecosystems are included among the vital functions provided by nature, since not all these services will be deemed necessary by everybody at any time. Secondly, whenever nature's intrinsic value is invoked as an argument against substitutability. Thirdly, whenever the technical limits to the latter are deemed absolute instead of historically contingent. These arguments are legitimate, but should not be presented as pre-given limits to nature's replacement beyond which sustainability cannot *work*. As Holland points out,

the concept of natural capital contains an epistemological variable: changes in
the level of natural capital are contingent, not upon changes in the natural world,
nor simply on its actual utility, but upon changes in assumptions about its utility
(Holland 1999: 61).

Furthermore, if we leave concerns about naturalness aside, it is conceivable that
further technological innovations may turn irreversible natural capital into fungible
or even disposable natural capital in the future: cloning animals is a case in point.
Hence the notion of critical natural capital could comprise a *diminishing* amount
of natural beings and processes. This is not to say that we must actually dispose
or replace them. Rather it is to confirm that substitutability is not an absolute
concept. And given that nobody defends – as of yet! – a complete replaceability
of natural capital, the defence of any version of sustainability is actually grounded
not so much in the present or future *feasibility* of a given socionatural balance
therein proposed, but rather on its *desirability* in relation to its alternatives. In
other words, it remains an essentially open decision.

Neither Weak nor Strong: Towards a Post-natural Sustainability

Sustainability can thus basically be either weak or strong, depending mostly on
how much natural capital is deemed replaceable by human-made capital: such
are the terms of the debate. Yet it is unlikely that these very terms, despite their
usefulness, are the most appropriate for reflecting on a post-natural sustainability.
For one thing, the humanisation of nature and its transformation into human
environment has entailed a deep process of hybridisation that renders the separation
between natural, human, and cultivated capital somehow unrealistic. By making
this sort of abstractions, a muddled reality characterised by interaction and mutual
contamination is conveniently stylised, but also misunderstood. Thus there exists
the delusion of an ethics of substitutability, or even the idea that a strong version
of sustainability is still within reach.

On the contrary, the socionatural metabolism has increasingly deepened the
mixture between the natural and the artificial, rendering the former notion less
and less meaningful. Although we tend to neglect it in the discussion about natural
capital, *cultivated* capital is probably the category that should be underlined as the
one that best captures the current state of socionatural relations. Human influences
upon natural beings and processes, either deliberate or indirect, are manifold.
They amount to a long process of hybridisation that blurs the frontiers between the
natural and the human – cultivated capital being precisely the intermediate product
of the resulting interchange.

By hybridisation I mean the environmental recombination that results after
humanly originated processes and artifacts have exerted a variable degree of
influence on natural beings and processes. It is not a new phenomenon, but it has
become more and more intense with the accumulation of scientific knowledge since

the beginning of Modernity. As the previous discussion on naturalness showed, human influence can be manifest or remain hidden – ranging from grafts or half-breed species to chemical processes affecting the soil or even, at its most spectacular, global warming. As already noted, it is not in vain that the term "anthropocene" has been coined to describe a period in human's history wherein humans exert a great influence on the planet's ecosystems (see Crutzen and Stoermer 2000). The outcome of this pervasive influence is a new environmental equilibrium, whose consequences are not so easily discerned. For instance, although biodiversity has decreased due to human action, an increase in the Earth's temperature could very well – it happened once – produce an eruption of ecological diversity.

On the other hand, this process of socionatural imbrication hints at the possibility that the notion of substitution may itself be flawed. Because it could well be that, properly speaking, there is no such thing as a nature-human *replacement* – but rather a mixture, an assimilation, a symbiosis between them. When we say that a manufactured-capital substitute requires inputs of natural capital, as we mentioned earlier, what kind of manufacturing processes are we describing? What sort of final products? Not certainly a whole artifact, but one made of both natural elements and human ingenuity. Let us think of books, computers, prosthesis. At the very end, *everything* is both natural and artifactual. It is in this regard that a full replacement must be considered an absurdity: there is no such thing. A different matter is the appearance of naturalness of the final product, which will depend on the human intention and the subsequent design of the thing or being or habitat in question. Are we trying to protect the *essence* of nature, the integrity of natural forms, or just their appearance of integrity?

After all, sustainability is not about society and the environment as separated realms: it concerns their *relations* in the long term. These relations are intrinsically dynamic, as we have already noted. But such dynamism is not just a social quality inflicted onto nature – it is also a natural quality. Otherwise there would barely have been a *change* in the course of socionatural history. This change is caused by an intricate network of influences, by which all kind of living beings are in contact with each other, pushing evolution forward. At a given moment, the human species becomes the most influential agent of change. But nature, the nature from which the human species itself stems from, must be alive and dynamic for that human influence to bear fruits –whatever the moral judgement is that we may express about the latter now.

Herein lies a typical weakness of classical environmentalism. Nature is not a passive object of human action, but rather a dynamic entity – consisting of endless entities – that changes in its own and changes in contact with humanity. Seen in this light, the notion of natural capital seems unable to portray this crucial feature of the natural world. As Biesecker and Hofmeister (2006, 2009) have been trying to underline, nature *lives* and is in itself *productive*, so that we should not see it just as a limit to human economic activity, but rather as forming a non-separable unity of productivity and re-productivity with the latter:

> *The productivity of nature is at the same time re-productivity.* (…) The production
> system "Nature" is then simultaneously starting point (productivity) and
> outcome (product) of the process of (re)production (Biesecker and Hofmeister
> 2009: 183).

In this context, hybrids are *processes* that communicate society and nature as
well as *products* that are nature-culture products – the latter being the outcome
of the former. As Marx already pointed out, this mixture is the very condition for
human life, as expressed in the notion of the socionatural metabolism. In the end,
everything is bound to become an hybrid: human influence cannot help but being
all-pervasive. We could then say that everything will become *human*, but in a
general sense the reverse is also true: everything remains natural. Yet what are the
implications of this for sustainability?

To begin with, although the notion of natural capital may retain its usefulness,
its usage must reflect the fact that naturalness is not an absolute category and thus
take into account the general process of hybridisation between society and nature.
Instead of focusing on natural capital as such, an extended notion of cultivated
capital should be at the centre of the analytical stage. As for critical natural capital,
it would arguably be reckless to discard a notion with such symbolic force when
it comes to enact limits to human activity. At the same time, though, the essential
variability of critical natural capital *due to* social influence and manipulation
should be properly acknowledged. After all, it is critical what appears to be critical,
until it ceases to be so. Besides, critical natural processes do not need to remain
untouched to provide key services to mankind. They can be altered, amended,
intervened – thus remaining critical without remaining fully natural. It is plain to
see that a strict distinction between nature and society is untenable.

On the other hand, a post-natural sustainability should be freed from
the constrictions posed by the strong interpretations of natural capital and
substitutability. Nature's dynamic and productive quality turns any steady-state
view of sustainability into a somewhat unrealistic choice. Yes: still a choice. But it
is not a necessity. By setting nature as an independent variable in the socionatural
equation, strong versions of sustainability bizarrely adhere to a delusion – the
delusion of the still photography of nature. Interestingly, this mirage is also
recurrent in the idyllic visions of a remote pastoral nature. Such view leads to
a static calculation of natural goods and resources, whose amount is to be kept
stable via a severe restriction of human activity. As if that human activity were
not characterised by an intermingling with nature – through a general process of
hybridisation – in the first place.

Restrictions of this sort are traceable in a number of tools usually employed in
sustainability studies. For instance, the morally grounded duty to maintain a non-
declining level of ecological and social indicators for future generations can easily
become a theoretical trap, insofar as a self-imposed future constriction dictates
not touching a number of natural assets. This reasoning implies that, were social
activity to continue, the overall welfare would be halted or diminished. But why?

It could be increased as well, as it has done consistently throughout history. Rather than focusing on a passive nature exhausted by humanity, socionatural exchange and interchange ought to be emphasised.

The same goes for sustainability indicators as prominent as the ecological footprint (Wackernagel and Rees 1996) or the environmental space (Opschoor and Weterings 1994). Respectively they try to measure the overall amount of environmental resources that are used to produce a given good and the corresponding share of the planet that may be used without depriving future generations of the resources they may need. Their usefulness lies in their ability to approximately measure the impact of ongoing human practices, but therein lies their considerable limitation as well. Let us take this definition: "The ecological footprint concept can be defined as the total area required to *indefinitely* sustain a given population at the *current* standard of living and at an *average* per capita consumption rate" (Moffat 2007: 329–30; my emphasis). It is indeed a still photograph, since it disregards or fails to take into sufficient account decisive factors such as future technological progress or changes in social preferences (see Ponthiere 2009; Ayres 2000). The entire socionatural history has been on the contrary marked by the apparition of the unexpected: new usages, discoveries, reactions. In fact, there is a hint of Malthusianism in these indicators: "A major objective is to estimate the relative share of global resources appropriated by a certain human population, activity, region or country as a basis for decision-making on (un)sustainability" (Deutsch, Folke and Skanberg 2003). If the ensuing decision is to impose a pre-given corset on human-nature exchanges, sustainability may in fact be severely hindered rather than enabled: low substitutability restricts the system's ability to provide for human well-being (Ruta and Hamilton 2007: 48).

Thus a viable post-natural sustainability should also openly accept the role of science and technology as major facilitators of environmental adaptation. It is through science and technology that we are able to measure society's ecological impact and it is through them that such impact can be diminished. Moreover, it is they that make a control of the increasing hybridisation process feasible. This does not add up to a mere continuation of past technological practices. In fact, it is the opposite: "Our management of future technology acceleration has to reverse the effects of past technology acceleration" (Brand 2009: 19–20). Although it would be reasonable to suspect such a dramatic turn in the human uses of technology, this is a gradual evolution rather than a rupture, a part of a slow process of cultural re-orientation whose final destination remains yet – as it must – uncertain. Environmentalism has never really come to terms with a scientific-technological solution to sustainability, probably because of the fact that such constellation is deemed by greens as the main culprit of environmental deterioration, which makes it either unbelievable or certainly unacceptable for them that it could provide as well a solution to the latter. It would also be a solution more prone to substitutability than the much-preferred strong version of sustainability. Hence the cliché that technology is not the solution. But why not? Just because we do not *wish* it to be? Technology will not by itself determine the amount of natural forms that will be protected, but it can help to make hybridisation sustainable – not to mention the fact that a wealthy sustainable society, as we shall defend, is more able to protect nature than a poor one.

Still, a question remains concerning the amount of nature that is to be protected *aside* from those natural units and processes that in each historical moment are considered irreplaceable in light of their contribution to basic environmental functions. Sustainability is not primarily concerned with nature's protection, as has been already noted. It is, after all, an essentially anthropocentric endeavour (Norton 2005). Yet the question of nature – what to do exactly with it – lies at its very centre. How much nature *must* we protect (in order to protect ourselves) and how much nature we *wish* to protect (for its own sake or on behalf of other human interests aside from survival) are the two guiding questions for defining this dimension of sustainability. The latter contains a strong technical content, whereas the former rests solely upon value judgements. In fact, the consensus about the nature we *need* sets a limit for the decision on the nature we *wish*.

That may sound incongruous. Protecting nature after the end of nature? Indeed. Neither naturalness nor hybridisation are absolute categories. On the contrary, they are gradable – depending on the degree of human influence exerted upon each biological process, natural good, or ecosystem. In this regard, although it is hard to think of untouched natural environments, there are indeed natural units or ecosystems that have not, or have barely been, transformed by humanity. Furthermore, there are variable degrees of natural appearance, corresponding to different degrees of an hybridisation that may not be visible: nature in a genetic sense is not the same as nature in a qualitative sense (see Birnbacher 2006). Henceforth three different subjects of protection – and thus contention – can be distinguished:

1. *environmental functions*, which may or may not require the preservation of an ecosystem integrity, depending on the likeliness and extent of their substitution;
2. *nature in a genetic sense*, namely, the historically grounded integrity of living beings, biological processes and ecosystems upon which humanity has not or barely or just less intensely influenced;
3. *nature in a qualitative sense*, that is to say, natural forms which retain an appearance of naturalness irrespective of the real influence exerted on them by human beings.

The discussion is then bound to take place on many different levels. The radical openness of the sustainability principle encourages such discussion, because the answers to both questions – how much nature we need and how much we wish for – are never closed. On the one hand, changes in scientific knowledge and technological ability may render some given limits to substitutability obsolete. On the other hand, social preferences may change too, altering our view of socionatural relations and hence modifying our decision about the degree of nature's protection.

This latter discussion should not be mixed up with the former, as is the case whenever nature's amenity functions (serving humans' moral, aesthetic, scientific interests) are included among strictly environmental functions.

In sum, a post-natural sustainability is neither weak nor strong. Insofar as nature has been transformed into human environment after a long process of hybridisation, it does not take for granted the validity of notions such as natural capital or substitutability. Nature and society are intermingled, the former being productive rather than static: a dynamic element of social activity and human economy. Yet nature's question must be answered – even after the end of nature. But that very question about nature is to be discussed in a framework that is preferably open to substitution and hybridisation, while constantly keeping in question the degree of protection granted to nature.

Chapter 5
The Politics of Sustainability

The Search for a Sustainable Polity

After theory, comes politics. It has been suggested so far that sustainability is an inherently open principle, which does not have a given content attached to it, hence leaving the shape of the sustainable society indeterminate. However, it should be underlined that a minimum set of standards must still be respected, so that the public debate and practice of sustainability takes place upon an ecologically sound foundation. But apart from that, the outlines of sustainability, including the degree of protection for nature or the emphasis on social justice, are left open. So far, so good.

Yet is this a realistic account of the politics of sustainability? Is real green truly realistic? Because a sustainable society that leaves its content indeterminate surely looks like it is a paradoxical – even contradictory – one. It is legitimate to ask how such scheme is going to be implemented. Furthermore, a key question arises as to whether this view of sustainability is going to provide the adequate outcomes. Insofar as sustainability must be, first and foremost, actually sustainable, is there any guarantee of that happening? Not to mention the fact that no clear justification has yet been offered as to *why* sustainability should be embraced as a comprehensive social goal.

Is sustainability compatible with democracy? What kinds of sustainable society are possible? Does sustainability determine or condition social organisation? Such are the key themes for any exploration of the politics *of* sustainability. Or, in other words, such are the questions that must be cleared up in order to decide which politics *for* sustainability is to be chosen. The ensuing framework will constitute the means for socially arranging the general principle of sustainability. Thus the challenge consists in finding a political articulation for it that is both effective and open – namely, sustainable and democratic. The search for that framework will be dealt with gradually in this chapter, as well as in the next section.

An obvious starting point here concerns the relationship between sustainability, environmentalism and democracy. After all, we are not concerned with just any sustainable society, but with a democratic one. Thus it is the democratic politics of sustainability that must be taken into consideration. And it is here that a crucial tension arises, since it makes no sense to invoke sustainability as a social guiding principle without taking its effectiveness into account: sustainability must work! Yet can democracy provide this? Or are we doomed to perish due to the inability of our political system to implement the necessary reforms? Idyllic images of the future aside, an ecological re-orientation of society cannot be achieved without certain societal changes that are bound to provoke resistance and protest. Not everybody wishes to have a windmill in the backyard. Henceforth, a peaceful transition to sustainability can only be conceived out of naivety – unless we choose a political framework that translates the inherent openness of the sustainability principle into the social realm without sacrificing its effectiveness. Such is the challenge.

Admittedly, coming back to the well-known tensions between environmentalism and democracy may seem superfluous, all the more so since the latter remains in practice firmly engaged to the latter. But this pledge is misleading, for a number of reasons. Firstly, because the foreseeable rise of climatic fears may well reinforce the temptation to question democracy as the best suited political means to mitigate global warming. Secondly, this very ecological urgency is bound to exert a considerable pressure onto the democratic *and* liberal foundations of Western societies, which may in turn influence both the way in which we see sustainability (less of an open principle and more a pre-given plan) and the political arrangements we choose to pursue it (making them less liberal-democratic). Thirdly, this relationship is also important in connection to the reframing of environmentalism that this book advocates. Renewing green politics also means a full acceptance of sustainability's open quality, which in turn requires the previous normalisation of environmentalism's relationship to democracy – and, as we shall see later, liberalism – in the normative realm.

To put it differently, it is not enough to claim the preferability of democracy over other political regimes. Sustainability itself must be democratically conceived, for society to remain democratic in the quest for sustainability. Before politics, it seems, more theory is required.

Environmentalism against Democracy?

The relationship between environmentalism and democracy has traditionally been characterised by a recalcitrant ambiguity. Certainly, green theorists and activists have adhered for years to democratic procedures, thus dispelling the authoritarian flavour of some green strains especially active during the seventies. But that is not enough to make the grounds of such ambiguity disappear, namely: the fact that democratic *procedures* cannot possibly guarantee green *outcomes*. This is the heart of the matter, such is the root of the green mistrust of democracy – and not without reason. An open decision-making procedure is bound to frustrate whoever is seeking a particular result from it. Henceforth, the more clearly is environmentalism defined by a moral consequentialism which demands the realisation of some substantive goals, the greater the clash between ecological and political values will be. After all, a morally indisputable end – like, say, the protection of nature – is not politically negotiable. Thus the paradoxical result of nature's politicisation is but another form of de-politicisation. Why? Because if nature is described as a critical problem requiring exceptional solutions with pre-fixed outcomes, there seems to be no room for a necessary connection between sustainability and democracy.

To explain this conflict, it will suffice for now to concentrate on a procedural conception of democracy, understanding it as a collective decision-making device based on the majority rule. According to this minimalistic view, a democratic procedure is a method of determining the content of legally binding decisions, so that the preferences of the citizens have some formal connection with the outcome in which each citizen counts equally (Barry 1979: 156). Political equality, effective participation, a sufficient understanding on the part of citizens, control of the agenda, enough participative inclusion: the fulfilling of these criteria make a full procedural democracy in relation to its *demos* feasible (Dahl 1979). Admittedly, a procedural democracy entails a priority of means over ends (Bealey 1988: 21). But then again, the procedure itself could not be set up nor function without some underlying principles – among them equity, rationality, tolerance, and liberty. At the same time, though, it is hard to deny that any conception of democracy is in itself procedural. It is a collective process of communication and decision-making involving, formally and informally, both citizens and institutions. The process may adopt different shapes depending in turn on different *conceptions* of democracy – ranging from representative to participatory, from direct to deliberative. But the conflict between environmentalism and democracy comes up earlier. Robert Goodin presents it very clearly:

> To advocate democracy is to advocate procedures, to advocate environmentalism is to advocate substantive outcomes: what guarantee can we have that the former procedures will yield the latter sort of outcomes? (Goodin 1992: 168).

The answer is: none. If democratic rules are respected, there is no way to assure that greens will achieve the results they aim for. Let us summarise them for now as the achievement of sustainability. The problem is that sustainability and democracy clash because the latter cannot guarantee the realisation of the former. Henceforth, environmentalism clashes with democracy too. But, it does so whenever it defends a conception of the good that is not open to negotiation or compromise – a metaphysical conception of nature that demands a closed conception of sustainability.

Even though as important a thinker as Robyn Eckersley (1992, 2002) has claimed that democracy is the most appropriate political form for implementing ecocentrism, a flagrant contradiction remains between the democratic procedure and any conception of the good that demands a thorough social transformation to be fulfilled. Such is arguably the case with ecocentrism itself or any other deep strain of environmentalism, as well as with the demand for a strong version of sustainability. That is to say, a democracy *is not supposed* to produce green outcomes in the absence of a sufficiently strong cultural consensus. Ironically, the kind of sudden consensus being currently forged under global warming's pressure may not serve green traditional goals either – insofar as it is a survival-orientated, instead of a nature-orientated, one.

Henceforth, a most solid link between environmentalism and democracy seems to be hindered by the core values of the former. A tension seems to exist between some green values and some democratic values, which is manifest in the relationship between sustainability and democracy. Whenever environmentalism defends the intrinsic value of nature, or any other morally strong commitment to nature's protection, it is using a language that is incommensurable with that of democracy. Is democracy capable of protecting nature? Brian Baxter does not think so: "it is clear that just leaving the defence of the moral claims of the non-human to the normal processes of human democracy will be inadequate" (1999: 112). Democracy is all too human! Its moral shortcomings are obvious from a green perspective:

> To investigate the relationship between democracy and nature, then, is to touch upon an old and fundamental concern about the moral integrity of democracy in new but sometimes surprisingly recognizable ways. (...) [It] is as though the natural environment has come to embody, through its deterioration and destruction at the hands of democratic society, the very flaws critics have always assumed to be endemic to democracy (Pepperman Taylor and Minteer 2002: 5).

Thus green values end up being *external* to democratic principles – or vice versa. And democracy itself runs the risk of becoming superfluous on the face of the ecological challenge, an option subordinate to the main goal: realising the green conception of the good. In this view, democracy has nothing to do with the most *legitimate* decision, but with the *best* outcome, namely, the *right* outcome as defined by a given set of participants – greens themselves. A sort of visionary ecocentrism comes to the fore here, which claims to possess a certainty that justifies an exclusive legitimacy for decision. For instance, Laura Westra suggests that we should replace the "sacred cow" of democracy, incapable as it is to solve the ecological crisis, with a platonic philosopher-king (Westra 1993). And such rationale is not new.

In fact, this anti-democratic temptation was once strong enough to produce a whole strain of thought: the aptly called eco-authoritarianism that flourished some 30 years ago. In a way, this proposal is representative of that time, a decade in which ecological fear was widespread and collapse seemed round the corner. Yet it is also the expression of a latent green instinct, more prone to come up to the surface in times of ecological unease – just like, by means of global warming, ours.

It was William Ophuls who gave a more precise shape to previous contributions by Garret Hardin, Paul Ehrlich, and Robert Heilbroner. Ophuls argued that the natural condition of the civilised human being is the *dearth* of the resources upon which his existence depends. Distributing scarce resources in an orderly way is, in fact, the very foundation of politics. Whenever ecological scarcity manifests itself, the survival of the community becomes – once again – the basic political problem. This means that luxuries such as democracy and individual liberty must be suspended, as their conditions of possibility disappear. Democracy is not viable under the pressure exerted by this pre-political problem. The influence of the well-known "tragedy of the commons" (Hardin 1977a) is evident. Scarce, collective resources must be protected and regulated to avoid an ecological calamity. This protection can only be enacted by political institutions exerting a sufficient degree of coertion, a pattern that entails the possibility that democracy does not survive (Ophuls 1977: 152; Heilbroner 1975: 86–90). A green Leviathan is thus suggested: "Only a government possessing great powers to regulate individual behavior in the ecological common interest can deal effectively with the tragedy of the commons" (Ophuls 1977: 154). This strong, centralised state is ruled by "if not a class of ecological guardians, then at least a class of ecological mandarins who possess the esoteric knowledge needed to run it well" (Ophuls 1977: 163). Therefore, a direct link is drawn between the technical nature of environmental problems and the technocratic rule. A green enlightened despotism. It is a remarkable path from cultural utopia to political dystopia.

A few years ago, these ideas would have just seemed an historical curiosity. But now that global warming is restoring a climate of fear very akin to that of the seventies, that curiosity is turning into a regularity, as the limitation of democratic regimes to rapidly impose the structural changes deemed necessary become evident. It is thus suggested that "humanity will have to trade its liberty to live as it wishes in favour of a system where survival is paramount" (Shearman and Smith 2007: 4). Actually, it is not only environmentalism that flirts with the green Leviathan. Take this passage from leading journalist Thomas Friedman:

> One-party autocracy certainly has its drawbacks. But when it is led by a reasonably enlightened group of people, as China is today, it can also have great advantages. That one party can just impose the politically difficult but critically important policies needed to move a society forward in the 21st century (Friedman 2009).

The coincidences are striking. A new version of ecological scarcity, the same authoritarian solution: not the philosopher, but the environmentalist-king. The following rule seems to apply: authoritarian solutions to environmental problems come to the fore in a way that is directly proportional to the rise of public fears about the seriousness of the latter. But fear is not the only explanation to the continuous attraction exerted by eco-authoritarian solutions. A deeper reason lies in some features of environmentalism itself – features largely and not coincidentally inherited from classical environmentalism.

On the one hand, the ghost of eco-authoritarianism is plainly Malthusian. A rigid view of ecological limits underlies the very idea of an unavoidable ecological scarcity. Yet, as has already been suggested, natural limits depends largely on socio-economic conditions. Nature does not produce a pre-given amount of resources, rather its supposed limits can be overcome – or negotiated – by human ingenuity.

Still, what the eco-authoritarian argument claims is that a centralised state that coerces civil society is more effective than democratic regimes when it comes to ecological sustainability (see Orr and Hill in Radcliffe 2000: 28). However, this is absurd. If the entire society must cooperate to reach a common goal, legitimacy cannot be separated from effectiveness (Saward 1998: 175). All the more since this model depends on social restraint: people must cooperate. Such is the first problem of the authoritarian solution.

A further issue concerns certainty. A green Leviathan is supposed to possess the technical knowledge needed to *impose* sustainability. Politics, understood as a mixture of conflict and dialogue, is replaced by science on account of a state of ecological exception. It is thus assumed that there is a single formula for sustainability: a steady-state economy for a severely austere society. Yet this is a gross oversimplification. Sustainability is an open principle that admits multiple developments, whose continuous process of implementation will actually entail the simultaneous adoption of a number of different solutions – as long as that process takes place in a democratic society or at least in a society where individuals keep a sufficient degree of personal freedom. The discussion on sustainability cannot be so easily settled. Furthermore, not even a zero-growth, authoritarian solution is easy to implement, since not even among ecological despots can consensus be taken for granted. Are birth control plans necessary? Can we eat animals? What should we do with all those planes that do not fly anymore? And so on, and so forth.

On the other hand, eco-authoritarianism is but another example of the paradoxical de-politicisation that a significant part of environmentalism has been applying to nature. It is a paradox insofar as environmentalism is supposed to pursue the reverse goal: turning nature into a political issue. Of course, it has done so. But at the same time, many green arguments push towards the opposite direction, only in a more subtle way. How? It is a simple reasoning. Firstly: if the ecological crisis is not a political problem, it cannot be politically solved, even less democratically. Therefore: if the ecological crisis belongs to the domains of either technical or spiritual knowledge, the path to the sustainable society is also to be signalled by those who possess such technical or mystical ability. Thus:

a. either the ecological crisis can be reduced to a technical problem, so that both personal rights and the majority rule are deemed obstacles for achieving sustainability;
b. or the ecological crisis can be reduced to a spiritual problem, whose solution lies outside democratic deliberation and decision-making.

Both ways, nature is de-politicised. It cannot be discussed nor negotiated. Therefore, we can speak of an original act of politicisation that brings nature into the political realm, but takes it out from the latter again. Insofar as this de-politicisation remains in place, it is difficult to see how environmentalism and democracy can be firmly linked. As Terence Ball wrote: "There is no logically or conceptually *necessary* connection between a commitment to the natural environment and a commitment to democracy" (Ball 2006: 132). Indeed. Such commitment could be said to work in practice, but not in theory.

This mismatch is a logical derivation from green consequentialism. If the protection of nature is a non-negotiable moral imperative, a decision orientated to such end can be *right* even against people's preferences. Therefore, a green program based on the intrinsic value of nature or in a strong view of sustainability discards the possibility of negotiation in a pluralistic framework: ecological values are beyond, hence also above, political values. Green consequentialists

> cling to a single moral "end" (intrinsic natural value) –an end advanced prior to moral deliberation and cooperative social inquiry– and attempt to anoint it as the *sole* principled ground for defending our environmental decisions policy choices (Minteer 2002: 45).

But is this not just the political translation of the famous ethical maxim suggested by Aldo Leopold? This said:

> A thing is right when it tends to preserve the integrity, stability and beauty of the biotic community. It is wrong when it tends otherwise (Leopold 1987: 224–5).

Correspondingly, a decision is legitimate not when it stems from a democratic procedure, nor when adopted in accordance to democratic values, but only when it respects such ecocentric mandate. Thus green normative foundations tend to displace democracy, lest an ecologically wrong decision is taken. Interestingly, this logic may easily lead to some form of eco-authoritarianism if left to its own devices, because the "things" that can cause harm to the biotic community are not only political *decisions*, but also a potentially infinite number of *actions* that are not linked to any formal decision-making processes. Democratic laws do not regulate every aspect of our lives. Hence the biotic community – a very wide concept – would still be threatened in an ecocentrically orientated democratic setting. It seems that only a static, authoritarian society can be properly sustainable. So much for democracy, then?

Reconciling Environmentalism with Democracy

This seeming incompatibility should not be dramatised. It is neither a problem for green activism, nor a practical problem that has to be dealt with on time for the overcoming of liberal democracy – since the latter is not happening overnight, and it may even be the case that it does not happen at all. On the other hand, it could be argued that these authoritarian arguments are themselves strategic claims within the green debate, that is, claims whose intention is to point out the intrinsic weakness of democracy in the face of systemic challenges rather than sketching a desirable society or actually imposing it. Yet this conflict, rhetorical as it may be, is not irrelevant either. It influences the direction of green politics, hindering its renewal, as much as it makes the search for a democratic sustainable society more difficult. Therefore, it does situate environmentalism in a quandary, forcing them silently to decide whether they want to be "first and foremost, environmentalists or, first and foremost, democrats (while admitting its many and varied weaknesses)" (Norton 2002: 23). They cannot be both simultaneously – unless they learn to be greens in a different way.

To be sure, attempts have been made to reconcile green and democratic values. However, insofar as there is no necessary or plausible connection between them, such attempts may even be seen as a case of wishful thinking (Humphrey 2004: 116). They leave a similar impression to that of the old theological debates about God's existence: the arguments are irreproachable, but ultimately unconvincing. Such efforts, however, deserve our attention. Three main arguments have been advanced.

1. *The preconditional argument.* It suggests that, despite its procedural quality, democracy is also committed to those conditions that make the procedure possible in the first place. Among them is not only rationality or a disposition toward dialogue, but environmental conditions as well. Some green goals may then be included among the democratic means, insofar as the integrity of ecological systems should be considered a *generalisable interest* excluded from the argumentative dispute (Saward 1996; Dryzek 1994). This is true. However, granting the ecological subsistence of a democratic regime is not the same as preserving the integrity of the biotic community. There are many ways of being sustainable and a good number of them would be deemed unacceptable by environmentalism for being not protective enough of natural beings and processes. In short, sustainability may be adopted as a general principle, but it still has to be discussed and negotiated. On the other hand, ecological sustainability is the pre-condition for any political regime, not just for democracy. Therefore, although sustainability can be made compatible with democracy in this way, the same does not apply to environmentalism itself.

2. *The value-emergence argument.* It claims that there is a link between the openness of democratic procedures and the objective validity of green values. Given that democracy has traditionally been able to regulate value conflicts in a reasonable way, it is to be expected that green values will ultimately emerge in such a framework, all the more in a participative rather than a liberal model of democracy (Hayward 1998: 161; Dobson 1996: 139). In other words: democracy, if authentic, is bound to be green. But is it? The alleged *objective* superiority of green arguments must not be mistaken with the validity that is *intersubjectively* granted to them. Not even in the supposedly ideal speech situation facilitated by a deliberative framework is there any guarantee that the *right* arguments will be chosen. After all, the

recognition that something is morally good does not automatically entail giving preference to it: people can choose their interests over their ethical intuitions in spite of their morally diminished status (De-Shalit 2000: 138). Eventually, this argument turns out to be instrumental: democracy will be necessary for environmentalism *if* green values actually emerge in the course of democratic procedures, but if that fails to happen, well, that is democracy's problem rather than environmentalism's.

3. *The deontological argument.* In this case, the premise is the democratic commitment to include every relevant interest in the political process. This inclusion may adopt two forms. On the one hand, changes in the democratic procedure can be made, so that the interests of an enlarged moral community, including the natural world, are represented in the political realm (Mills 1996). Thus some outcomes are expected, but not directly prescribed. On the other hand, the recognition of that enlarged community can be implemented via an adscription of rights that is grounded on the liberal principle of autonomy – be it recognising that the preconditions of human autonomy include not only legal and material pre-conditions, but also ecological ones, or be it recognising that the natural world itself deserves the enjoyment of its own autonomy (Dobson 1996: 142–5; Eckersley 1996: 214). Democracy's own logic would then lead to a necessary connection to green substantive goals. Yet it rather seems that a hypothetical outcome of the political process – enlarging the moral community – is disguised here as either a procedural modification or a mere correction of the autonomy principle. These are substantive outcomes: ends in themselves rather than simple means. As such, they cannot be excluded from a *previous* democratic decision.

There is nothing wrong with this set of arguments. They offer valuable ideas that may help to improve the democratic polity. Yet they must themselves be democratically considered, weighed, before they are adopted, if adopted at all. Thus they do not provide *in advance* a necessary connection between environmentalism and democracy.

But it could not be otherwise. Environmentalism is not going to be anchored to democracy by the compulsory greening of democratic institutions, nor by making democracy writ large less liberal and more participative. Rather it is environmentalism itself that must be transformed, if it is to be fully reconciled to democratic institutions and procedures, moreover, to a liberal-democratic society. An obvious alternative is not doing so, that is, keeping environmentalism as an outcome-orientated ideology that tries to find its way into democratic procedures but is perpetually frustrated by the latter's limitations. However, although environmentalism may not need democracy, democracy certainly needs environmentalism – yet a different, rejuvenated one. The case for reframing green politics is thus reinforced.

First and foremost, such reinvigoration entails putting an end to the exaggerated influence still exerted by classical environmentalism in the axiological architecture of green politics. The green epistemological essentialism that tends to derive moral mandates directly from nature should be replaced by a more realistic view of nature and hence of the political possibilities attached to it. Reframing is thus also displacing – the new green politics is opposed to the old green ideology. In other words, environmentalism must free itself from nature. Only thus can it open itself to contingency, pluralism, and politics. And only thus can a truly modern green politics be developed, one that is able to make a difference – a green difference! – to liberal democracy.

It is arguably the influence of classical environmentalism that has pushed green thought into a restricted conception of politics: as the simple epiphenomenon of an ecocentric ethics. If the green view of nature is adopted as a source of moral goals, moreover, as a blueprint for arranging social order, politics will be completely removed. The reason is that the foundation of politics is thus situated outside politics: a certain view of nature and of our relationship to nature. More exactly, this green ontology is situated *before* politics, so that the task of the latter is precisely to put this view in practice – as opposed to discussing or weighing it up against other goals. It is the relative autonomy of politics regarding morals that disappears. Politics becomes a mere tool for reproducing the natural order within the social order. And there is a connection between the refusal to see the social and historical embeddedness of nature and the naturalisation of the moral and political reflection. Nature is de-politicised, politics is de-naturalised.

This is especially the case whenever environmentalists are inclined to normatively translate nature's features into social postulates – whether to guide individual human action or social organisation. A general compatibility between political and natural goods is assumed by green philosophers that allows a reciprocal translation (Jayal 2001: 65). Hence nature is read with the help of ecology and a number of its features are translated into moral or political principles. Nature can hence become historian, physicist, systems builder and engineer, careful mother, teacher, philosopher, theologian or political scientist (see Caes 1995). And if the natural world is seen as a web of relations, then the self is also a "relational self", as much as society must be organised according to a "relational political model" based on the interrelation of biosocial communities (Mathews 1995b). Arguably the peak of this approach is the bioregionalist position, according to which the natural features of each territory – the bioregion – are to determine the sociopolitical shape of the human community settled in it: "a life-territory, a place defined by its life forms, its topography and its biota, rather than by human dictates; a region governed by nature, not legislature" (Sale 1985: 43). A quite Leopoldian purpose.

We can think of politics as the formal means by which the conflict arising from pluralism and scarcity is dealt with. Pluralism means that people possess a different conception of the good life and hence aim for different goods, whereas scarcity, understood in a wide sense, entails that not all conceptions of the good can be realised at the same time. They cannot, because even though a scenario of material wealth can be anticipated, wherein nobody suffers a significant lack of resources, there would still remain conceptions of the good whose realisation depends on a radical transformation of society – e.g. Marxism or ecocentrism. On the other hand, politics is the *formal* means to deal with the ensuing conflict, but not the only means: the very functioning of societies constitute the *informal* means by which reality takes shape, via an infinite number of individual actions and decisions outside the political realm. We will come back to this later. In this context, then, morality is concerned with the desirable, whereas politics is engaged with the possible: not the other way around.

Seen in this light, the de-politicisation of nature amounts to a de-naturalisation of politics, since the only acceptable outcome of the former is a re-orientation of society in the direction of a strong version of sustainability, i.e. anti-politics. It is thus understandable that the moralisation of green political thought has induced not only a neglect of the political, but also an undermining of the former's vitality and flexibility (Torgerson 1999: 14). I am not suggesting that environmentalism cannot seek inspiration in nature, nor that it has to abandon the very purpose of protecting it. Yet nature cannot be *naturally* defended. And a good reason for that is that the socionatural relationship can adopt several shapes, not a single one which is deduced directly from nature itself. Another is the lack of realism that afflicts many green representations of nature, precisely now that nature has been replaced by the human environment. It follows from here that environmentalism should not be concerned as much with nature as with the environment, that is, with sustainability. Thus it is through sustainability that environmentalism can be necessarily linked to democracy.

Yet not through any particular conception of sustainability, but by defending instead the general principle of sustainability as the framework whereby different versions of it can be introduced, discussed, and pursued. It remains to be seen why a liberal-democratic society should embrace the goal of sustainability, a question which may be not so difficult to answer in practice as it is in theory – as we shall see in the next chapter. However, as we are dealing here with the relationship between environmentalism and democracy, in order to connect them in a necessary way, we must firstly resolve the reverse question: why democracy is good for sustainability? And the reason lies in the radical openness of this very principle. But not just because of the manifold sustainable societies that can be conceived of. Also in view of the fact that any version of sustainability is marked by some features – complexity, indeterminacy, uncertainty – that can only be properly managed by a liberal-democratic society. An open sustainability principle demands both an open polity and an open society. Such is the bet of a post-natural environmentalism.

Towards a Green Society: Open *versus* Closed Sustainability

So far, it has been suggested that sustainability is a general principle that admits several different versions, depending on the way in which it chooses to re-organise the socionatural relationship – which basically means how much nature it decides to protect and how much substitution of natural capital it allows. Yet this way of looking at the content of sustainability, summarised in the opposition between weak and strong sustainabilities, is not enough when the political articulation of this general principle is dealt with. The fact that we talk of a sustainable *society* indicates that the sociopolitical dimension of sustainability is as important as its ecological one. Moreover, the two are closely intertwined, because it is obvious that the way in which a society operates will determine the kind of socionatural equilibrium it reaches, or fails to reach.

However, the political organisation of sustainability is often overlooked in the typologies that try to make sense of the different possibilities implicit in such a far-reaching concept. Normally, they centre the analysis on what is to be sustained and why, how the ensuing change can be justified and by what means is it going to be implemented. But a properly sociopolitical dimension should be added that helps us to understand the relationship between sustainability and social organisation. Does sustainability restrict the shape of society, or the latter is flexible enough to accommodate several sustainabilities? What is the best way to organise society in order to achieve sustainability? Is that society democratic or authoritarian? What kind of citizen's participation is required?

I would like to suggest that a distinction can be made between two different ways of conceiving sustainability: as a general principle and as a pre-given state. That is, respectively, between an *open* and a *closed* account of sustainability. They are not so much particular theories of sustainability, as meta-theories about the way in which the latter should be approached. However, both include particular suggestions as to which kind of sustainable society is preferable, since a set of assumptions about the re-organisation of socionatural relations is implicit in their general approach to the subject. Interestingly, though, an open sustainability might end up developing the view of socionatural relations advocated by its closed antagonist, whereas the contrary is not, cannot be, possible.

Closed Sustainability

In this case, the content of sustainability, namely, the shape of social, cultural and economic relations with the environment is *technically* or *ideologically* determined, leading to a model removed from public debate. This conception of sustainability gives priority to the technical feasibility or the ideological coherence of the model over its democratic – or liberal-democratic – elucidation. Thus it is imbued with a strongly consequentialist orientation: the outcome takes precedence over the procedure. This leads *de facto* to the abolition of both politics (understood as the intersubjective realm for deliberation and negotiation) and society (as the practical realm wherein deliberation and negotiations take place in a non-formal way). A particular version of sustainability is implemented, without it being previously discussed: because there is nothing to discuss.

In this way, sustainability's openness is substituted by an objectification of the sustainable society's content that allows the implementation of public policies at every level. This means that the *politics* of sustainability becomes simple *management*. There is no place for political judgement, nor for spontaneous social action. Two variants of this view can be distinguished:

1. A technocratic model, wherein sustainability is approached as an administrative issue, a matter of designing the right policies via scientific assessment, and implementing them bureaucratically. Instrumental rationality leads the entire process. Allegedly there are no value conflicts, and hence no need for public debate. Such technocratic variant corresponds to the weakest interpretations of sustainability, a sort of utopian techno-liberal model of development. It is business as usual. But it is also implicit in the eco-authoritarian model based on an ecologically enlightened Leviathan.
2. An ecotopian model, based on the restoration of socionatural harmony along the lines of a full ecologisation of society. The politics of such a sustainable society is but an ecological politics. While these ecological principles are objectively formulated, there is no need for public conflict. Of course, the kind of individual inhabiting it is assumed to be just the right one: cooperative, disinterested, a lover of nature. It is utopia as usual.

In sum, a closed view of sustainability gives precedence to the outcomes over the procedures – and even over the individual rights that underpin a democratic society. It assumes, in fact, no necessary link between sustainability and democracy. It can do so by either exaggerating the unsustainability of human society or underestimating the need for an ecological re-orientation. But either way it seems to depend largely on democracy's *absence*, that is, on not allowing any kind of public participation in the definition of sustainability's content. The fact that this closed model is not always explicitly advocated does not mean that it is not influential in our understanding – or misunderstanding – of sustainability. On the other hand, whenever green theorists or activists warn against the ecological crisis and subscribe to a radical overcoming of the current sociopolitical system, a closed sustainability is the next logical step.

Open Sustainability

Sustainability is instead conceived as a necessary societal goal whose content and conditions are nonetheless not pre-given. There is no single sustainability, but a whole range of different, even simultaneous, possibilities. The politics of sustainability are open to public deliberation and to the spontaneous development of ideas and practices. Sustainability is thus inherently open and radically democratic, since only a democratic society guarantees the necessary conditions for its achievement. Several reasons can be adduced in support of this statement.

Firstly, the indeterminacy of the future conditions of society – that is, of society's knowledge, preferences, and technological capabilities – makes it advisable not to close the socionatural relationship along the lines of a single understanding of sustainability. We do not know what it is that we shall know. Nor what we shall prefer! This not only has to do with particular uncertainties (affecting environmental problems or economic conditions), but also with a more general uncertainty concerning the evolution of human values and preferences. This uncertainty, in turn, is grounded on the *dynamic* quality of human societies – a key feature that is historically observable and suggests that a *static* projection for them simply does not work. Henceforth, sustainability must be pursued without betting on an irreversible model that restricts the general openness of the principle too much.

Secondly, although the role of experts and scientists is obviously crucial in the continuous research and experimentation that sustains a green society, there is no such thing as value-free scientific knowledge. Now, this is a trivial verification – almost no social scientist claim nowadays that science can be purely objective. But we must be cautious when treading this territory, because if we stress the constructivist side of science we may be also undermining its authority when it is needed. It might be best to state that there are several scientific *and* technological perspectives *and* solutions which are applicable to different versions of sustainability. There is, in short, no *single* scientific solution. It is true that there is not an infinite number of them, but not a shortage either. Still, the fact that the obstacles are sociopolitical rather than technical does not authorise a technocratic view of sustainability. An open sustainability re-politicises the socionatural relationship by encouraging a diversity of both morally-grounded conceptions of the good sustainable society *and* techno-scientific solutions for them.

Thirdly, then, the preference for a certain version of the sustainable society reflects a conception of the good – or at least a conception of the good green society. The conflict between different versions of sustainability is thus a value conflict. The general desirability of the sustainable society is one thing, the particular specification of it that is preferred *against* others is another. Society's moral pluralism – as well as the moral pluralism of different societies among themselves – is reflected in this realm too. Instead of an ideological conception of sustainability based on ecological economics or intrinsic value theory, value pluralism is to be accepted as the basis for sustainability (Norton 2005). Although this pluralism may not be self-evident, the increasing relevance of the socionatural choice will force people to elaborate, either more or less sophisticatedly, a preference. This preference is thus rooted in values as it is a normative matter. But it is also a political matter, insofar as it must be publicly discussed and experienced. For it to be possible, the general principle of sustainability must be adopted as a generalisable public goal, which in turn will operate as a pre-deliberative agreement upon which *different* versions of sustainability are defended (Grunwald 2009: 48).

Therefore, a democratic society provides the framework wherein an open principle of sustainability can operate because the latter is radically open to a society that gives shape to it gradually. And if we take into account the diversity, complexity and uncertainty that are inherent to sustainability, it is plain that a democratic organisation of society is the most appropriate frame for relentlessly discussing and implementing it.

> Indeed, uncertainty, combined with differing interests and perspectives, ensures that the quest for sustainability cannot comfortably be contained by the terms of technical discourse, but is pressed into a political context where the meaning of key terms is vigorously contested (Torgerson 1995: 11).

No single actor would be able to adopt each and every correct decision from a centralised position, nor is the richness of social actors and right-bearers to be artificially curtailed to supposedly simplify matters. The democratic process and the spontaneous processes of communication and action that take place within democratic society facilitate the emergence of ideas and solutions that should lead to a green sustainable society. On the other hand, the conflict of interests that surround many environmental problems are best handled in a democratic context, because the resulting decision will possess a greater legitimacy (Eder 1996: 206). However, it is obvious that there will always remain a certain degree of uncertainty about a democratically pursued sustainability. Nobody will be ever fully satisfied, insofar as the whole realisation of a single version of sustainability would require a cultural and axiological homogeneity that is hard to envisage. But then again, sustainability should probably be seen as many other normative principles, from liberty to equality and justice, that are aimed for but not – ever – completely realised.

Henceforth, the distinction between an open and a closed sustainability separates two different views of sustainability itself, rather than two detailed sets of prescription about the particular shape of a sustainable society. In this regard, the distinction is similar to that of Elliott (2007), who has distinguished between a *political* and a *metaphysical* conception of sustainability. The former provides an account of sustainability that relies not on controversial religious or philosophical views, but rather on fundamental ideas that all reasonable members of a democratic society can accept, whereas the latter does the opposite. We could then say that the adoption of a general mandate for sustainability on the part of society exhausts its metaphysical commitment – the rest is political. However, it is the case that an open sustainability provides a framework for discussing different versions of the latter that may be metaphysically inspired and hence "strong", whereas a closed sustainability does not leave room for any other version, unless it is a mere variation on itself. Open sustainability is then unmistakably liberal-democratic: it states that the sustainable society is not the result of heavy planning, but rather the unpredictable outcome of several layers of decision and practice in a great number of institutional, cultural and environmental contexts.

Yet it is obvious that both meta-conceptions hold particular assumptions about the kind of sustainable society that is preferable. In other words, they set pre-conditions that push society in a certain direction. This is more evident in the case of closed sustainability, since it restricts more strongly the variability that a sustainable society can admit. But it happens with open sustainability too. The difference lies in its openness, because the latter must admit the ongoing result of the debate and practice of sustainability, so that, were a strong version of sustainability to emerge as a result of them, it would be implemented, provided that it is compatible with the maintenance of a sufficient degree of flexibility and openness for the system at large.

But what are these assumptions? They can be summarised drawing on the following axis (Table 5.1). They should suffice, although it is obvious that some others could be added.

1. The first concerns the *moral philosophy* of these accounts, namely, their position regarding the value of nature and hence also their main justification for pursuing sustainability. Whereas a closed sustainability tends to recognise the intrinsic value of nature, or at least some variation of it, open sustainability is anthropocentrical, although that does *not* determine in advance the degree of protection to be granted to nature. The former sees the protection of the biotic community, humanity included, as the reason for going sustainable. The latter is more survival-orientated – as a general, encompassing reason that, again, does *not* exclude further or more sophisticated layers of justification.

2. The second describes *socionatural relations' guiding principle*, which is a complicated way of referring to the substitutability question. It is the guiding principle of socionatural metabolism, as it determines a more or less flexible relationship to the natural world. A closed view of sustainability imposes a high restriction for natural capital substitution and tries to protect the remaining naturalness as much as possible. An open view is more inclined to see the socionatural relation in terms of an increasing hybridisation and accepts *prima facie* a high degree of substitutability – that may or may not be adopted as a general standard.

3. The third indicates what the *practical philosophy* is of each conception. In other words, the general means by which the sustainable society is to be reached. There is a clear contrast here between a closed sustainability that promotes the radical transformation of our way of living and thus the abolition of the current socio-economic system, on the one hand, and an open sustainability that, by contrast, sees the gradual reform via ecological modernisation as the adequate means for arranging a green society. In this case, ecological modernisation could be replaced by other social tools if that were the result of the social debate on sustainability, but the general

commitment to reform likewise could not be replaced with a rupturist approach, since that would mean an irreversible change and an actual *closure* of a framework that is to remain sufficiently open to further re-definitions.

4. The fourth axis refers to the *sociopolitical dimension* of the sustainability quest. It signals the kind of particular arrangements that will constitute the political context of the would-be green society. The closed view of sustainability adheres to a finalist view, based on planning (whether it is centralised or de-centralised) and the introduction of environmental standards. The open view is dynamic and prefers a liberal-democratic setting – complemented by governance elements – as the sociopolitical means by which different ideas and practices of sustainability can be tested in a pluralist context.

Table 5.1   Closed and open sustainability

	Moral philosophy	Socionatural relationship's guiding principle	Practical philosophy	Sociopolitical dimension
Closed sustainability	Ecocentrically orientated/ Recognition of the intrinsic value	Naturalness/ Refusal of substitutability	Radical transformation of the current socioeconomical system	Finalist conception; planning/ introduction of strict environmental standards
Open sustainability	Anthropocentrically orientated/ Compatible with a variable degree of natural protection	Hybridisation/ *Prima facie* acceptance of a high degree of substitutability	Reform through ecological modernisation	Dynamic conception; liberal-democratic setting for discussion and experimentation

A conclusion follows, concerning the relationship between social organisation and sustainability: there is no pre-given shape for a sustainable society, but there are different possibilities that may be tested and chosen, many of them simultaneously, hence giving sustainability an organically evolving form. The lack of a single social shape for sustainability is a most relevant insight, since the opposite claim that sustainability *requires* a given form would otherwise render democracy expendable – or, at best, a matter of choice. Such would be the claim that a closed view of sustainability, in the foreseeable absence of social homogeneity, would be force to make: the social would subordinate itself to the ecological. It is only by opening the ecological to the political, as an open view of sustainability advocates, that we are able to simultaneously democratise sustainability and make democracy sustainable.

But, as we shall see, it is not compulsory to ground sustainability in a participatory version of democracy either, as green theorists so often claim (see Choucri 1999: 149). A democratic society is not the same as a radical democracy. And although citizen *cooperation* is certainly indispensable to a sustainable society, *political participation* is not. This is not to deny the possibility that a sustainable society *be* grounded on some kind of participatory democracy, rather to clarify that it *has* not to be so. In fact, it is easy to imagine that the sudden adoption of a direct democracy could be counterproductive to sustainability, insofar as the public opinion in its current state lacks worldwide the necessary competence to make constant decisions on complex environmental issues.

It could be argued that sustainability differs from other normative principles in a key aspect, namely, in the greater importance possessed by practical outcomes. Sustainability must work in a way that liberty or justice do not – they can be just partially fulfilled for the time being, as long as the promise of a more thorough realisation of particular ideals of liberty or justice can be achieved in the future. If a sustainability path fails though, there is no future. This is important, for three reasons. Firstly, any normative conception of sustainability must be empirically viable for it to be debatable in the first place. Some blueprints may simply be unfeasible, no matter how appealing they may be from a normative standpoint. In other words, the principle of sustainability is open, but not that open. Secondly, the conflict between different conceptions of sustainability will involve means as much as ends: the empirical feasibility of a given blueprint as much as the desirability of the societal model it portrays. Thirdly, disagreements are bound to appear *within* particular blueprints as well, concerning the means by which they are going to be practically implemented. Sustainability, in sum, is not an endlessly elastic principle – because it is not an endlessly elastic reality.

Now, this is true enough, but it might be less important than it seems once a general recognition of sustainability's need for a solid empirical foundation is made. In fact, such recognition may also be understood as being implicit in the principle itself. But the point is that separating normative ends from empirical means is not as easy as it appears in most cases when it comes to sustainability. To begin with, normative disagreements can be masked as empirical disagreements, when defenders of a particular blueprint try to discredit a rival version of sustainability. Besides, it is difficult to foresee whether a reasonably serious view of sustainability is going to become unsustainable, especially since means can be refined over time. On the other hand, given that sustainability should not be conceived as a final state, but rather as a dynamic process, the means by which we try to realise a particular conception of sustainability is also a constitutive part of it – empirical means acquire then a normative meaning. Finally, the debate between different versions of sustainability is not conducted primarily via formal argumentations in formal settings, rather it takes place in many different arenas and it is as much acted and lived as it is talked about.

Therefore, it is probably better to demand that every version of sustainability is grounded on a *basic feasilibity* upon which the debate can take place, so that obviously failed proposals remain out of consideration. How can we do that? Admittedly, this is not easy. But the role of the state and of the increasing number of institutions and organisations that evaluate the state of the environment and make projections about future developments of current trends should help to rationalise that debate. This might be seen as a loose solution. But an open view of sustainability is ultimately more advisable than a closed one even if we take this problem into account, since different sustainabilities that are not mutually exclusive can be pursued, thus avoiding that the empirical failure of a single blueprint can lead to a thorough social failure. Market mechanisms, for instance, would be more dangerous if they were the only means by which we pursue sustainability – but they are not, they co-exist with many other approaches and perspectives that in turn influence and refine them.

Sustainability's inherent openness is thus also a societal openness. Needless to say, this is the account that a renewed environmentalism should embrace: a fallible one that does not restrict in advance the outcome of the social process of experimentation and learning that the quest for sustainability consists of. The need to integrate climate change within sustainability's content makes this demand even more pressing.

Sustainability and Climate Change

The irruption of climate change as a major environmental issue in the public arena during the last decade, which has helped environmentalism to recover some of its force as an advocate of radical change, is also starting to strongly affect the debate around sustainability. That is hardly surprising, since no sustainable society is conceivable without tackling or at least minimising the impact of a phenomena as complex and encompassing as climate change. Although they still tend to be treated separately, mostly out of academic inertia, it is obvious that sustainability cannot be defined apart from climate change, and vice versa: "Dealing adequately with climate change requires the resolution of the climate change dilemma within the broader context of sustainable development" (Lawn 2010b: 116). Naturally, that climate change can be defined as a *dilemma* has to do with its peculiar and complicated character, which is in turn mirrored by the peculiar and complicated character of the decisions it seems to require to be made.

To be sure, there is no danger of understatement here. Climate change is said to represent the most globalised impact of humanity on nature in history, the complexity of the corresponding socionatural interactions leading us into "uncharted waters" (Lever-Tracy and Pittock 2010: 7). That makes it not only a very serious problem, but also a hardly manageable one, since the activities causing it are the very foundation of world economies and are not easily amenable to simple technological correctives (Dessler and Parson 2006: 2). Besides, it exhibits features that make it the implementation of unambiguous solutions particularly difficult: global and long-term nature, high uncertainties, multiple actors and only partially-formed remedial institutional mechanisms (Reyner and Okereke 2007). Emma Duncan has expressed it accurately: "It is a prisoner's dilemma, a free-rider problem and the tragedy of the commons all rolled into one" (Duncan 2009: 4). Needless to say, such a thorny combination has propelled green dystopianism again.

Thus frightening future scenarios are pictured. To mention just a couple of countless examples, Welzer (2008) sees a world marked by a rising inequality and full of climate wars, whereas Urry (2008) suggests that what lies ahead is either a Hobbesian conflict of all against all or an Orwellian digital panopticon exerting draconian social control. As usual, this bleak future is not unavoidable:

> We do not believe that collapse is an inevitable outcome; we merely argue that the evidence at the moment points to it as the most likely outcome and that social scientists should get to grips with this current reality (Leahy, Bowden and Threadgold 2010: 864–5).

In other words, it is the most likely scenario *unless* radical change is implemented. At this point, a familiar paradox appears. The call for radical change, in the name of a potentially catastrophic development of the planet's climate, stands in contrast to an appearance of normality in our everyday lives. Due to the abstract and long-term nature of the threat, it is hard for citizens to feel that such change is needed, in spite of the fact that they may *say* that it is and social scientists may even believe that they mean it (see Brechin 2010). It is not just that citizens may have grown relatively immune to green dystopianism after years of false alarms. Those that pay attention enough may be genuinely confused about what is going on. The *reality* and the *idea* of weather intermingle constantly. Should they believe that this time is different? Is climate change really a threat as formidable as it is portrayed? But what about those analyses that downplay the extent of the menace? Is it really hotter, or is it just that the media say it is?

In principle, public opinion should just rely on science – hence the activity of the Intergovernmental Panel on Climate Change as a bridge between science and the public. But then again, we have read Kuhn and Fereyabend: the sociology of scientific knowledge has convinced us that society is inside the laboratory and science can only reflect social priorities and political interests. How can we *just* rely on science? To some, actually, climatology is not saying the truth about global warming (Leroux 2005). Yet science must still be our standpoint, for there is no better alternative, even though it is a "post-normal science" whereby "facts are uncertain, values in dispute, stakes high and decision urgent" (Funtowicz and Ravetz 1993: 742). However, a misunderstanding should be avoided. It is in this context that Sheila Jasanoff (2007) has advocated the need to produce a more humble science, one that leaves room for ethics and renounces the modern dream of a complete control over nature. That is just about right. But the reflective re-shaping of socionatural relations, up to a point where we try to regulate the oscillations of the climate with our actions, is not precisely a *humble* goal, nor an absurd one, especially since there is no direct relation between the current scientific consensus and the green radical vision of a de-industrialised society. Although action must be taken, it should be a proportionate one. Devising public policies and fostering private behaviour as part of a climate change policy should not be used as a pretext for advancing a closed conception of sustainability. Sustainability must encompass climate change, instead of climate change simply closing up sustainability.

I would like to suggest that climate change's social dilemma resembles the one described by Blaise Pascal regarding God's existence. He famously reduced faith to a wager after considering the probabilities at stake. Pascal suggested that, although we cannot prove through reason that God exists, a person should bet on His existence, since living life accordingly one has everything to gain and nothing to lose, whereas, even more crucially, acting otherwise could mean losing everything and gaining eternal damnation (Pascal 1995: 123–5). Likewise, we do know that temperatures are rising, although we do not know how will they evolve in the future, while there exists the possibility that humans are an active agent in that process *and* they can still influence on it. Thus two related possibilities become meaningless: that humans have nothing to do with the climate's evolution or that they cannot influence the current process anymore. They become meaningless because we must maximise our chances, that is, we must act *as if* advancing towards sustainability could mitigate global warming or at least facilitating the least damaging adaptation to its effects. No other wager makes sense.

However, the need to *act* does not automatically indicate *how* to do so. Hence the public debate. We know that social engineering on a huge scale can fail miserably – as the twentieth century comes to show. Still, in the manner of a global insurance policy, a strategy for mitigation and adaptation is necessary. This strategy should be orientated to make possible the *continuity*, not the dismantling, of our current society. Neither a programme for ruralisation nor the low energy proposals aimed to scale back society into a network of self-sufficient communities are realistic (see Trainer 2010). They represent the comeback of green utopianism, although their usefulness in the debate of ideas should not be neglected: their defence of a *radical* transformation is necessary for achieving a *moderate* change. As Dyer writes:

> I like living in a high-energy civilisation, and I don't want to give it up. If it can be managed without causing a climate disaster, I would like everybody on the planet to live in wealthy societies that have the resources and the leisure to start looking after all citizens and not just the top dogs (Dyer 2008: 128).

That is why climate change should "work for us", as Hulme and Neufeldt (2010) put it. It should be used for improving our societies through reform, not to pursue an unfeasible rupture based on a miraculous radical change in people's values (see Hourdequin 2010). It is more probable that people will follow a given virtuous inertia than to expect a sudden moral epiphany that clashes brutally with contemporary lifestyles – lifestyles that, despite the contempt that social science tends to show, people may well *like*. Therefore, in a nutshell, it is unlikely that citizens abandon their smartphones in order to embrace the charms of a more embedded rural life. It will simply not happen, cynical as it may sound. It also may sound Panglossian, since many today do not have enough money to acquire a telephone and the sources of dissatisfaction remain plentiful. It is in this connection that radical perspectives, namely, those wishing for some radical changes in the current sociopolitical organisation, are to be seen as the legitimate expression of unmet needs and desires deserving attention. This is true for global warming as it is true for other social problems. Yet we should not make mistakes when considering the sources of change. It is unlikely that the latter can be provoked by a sudden moral realisation on the part of relatively affluent citizens – it is more probable that a gradual evolution will take place, influenced by a multiplicity of factors, moral as well as economic and technological. On the other hand, a reformist and gradual approach to social change does not preclude the possibility that radical changes are the final outcome of an emergentist rather than a revolutionary process. Thus we should do the possible within the reasonable.

But what does *that* mean? To begin with, it does not mean that the notion of sustainability presented so far has become invalidated. Unsurprisingly, classical environmentalists present climate change as the sudden and decisive proof that many old green positions happen to be right: nature is not abolished, human dominion of nature is not feasible, risks are everywhere. Therefore, we have been wrong and our worldview, together with our social organisation, must change. We cannot apply our old human solutions anymore:

> I am terrified by the hubris, the conceit, the arrogance implied by the words like "managing the planet' and 'stabilising the climate". (…) Why are we, with our magnificent brains, so easily seduced by technocratic totalitarianism? (Tennekes in Hulme 2009: 312).

However, we do not have any option other than trying to exert some degree of control over climate. After all, we find out what is going on with the climate because we try to exert such control (see Edwards 2010). Again, the latter should not be understood as a complete dominion, but rather as a sufficient, self-aware one. Mitigation policies are an attempt to influence climate – but I cannot see any arrogance in them. Furthermore, that we are able to discuss and devise strategies in the face of an abstract scientifically predicted threat should not be seen as a failure, but rather as a triumph of human reason. Similarly, the idea of an anthropogenic climate change does not demonstrate that nature has not ended, but rather comes to confirm in an unprecedented scale the merging of nature and society *into* the environment. As Leigh Glover puts it, "there is nothing natural left in the global atmosphere; humanity lives in and breathes an atmosphere that's an artifice of industrial activity and, consequently, the global climate is also now beyond nature" (Glover 2006: 254). If anything, climate change reinforces the case for a realistic sustainability.

However, crucially, an advantage of climate change in this regard is that the kind of measures it demands – mitigation and adaptation in a wide scale – should help to push the sustainability debate in the right direction. The reason is threefold. Firstly, climate change stresses by its very nature the issue of well-being and quality of life as much as that of pure survival. As the Hartwell Group (2010) has underlined, climate change is not so much a problem to be solved, as a condition to live and cope with. Thus we should take advantage of the changes it demands in order to live *better*. That is, in healthier urban environments, in knowledge-based economies, with the best public education and health care for all (see Baker 2006: 3). Thus sustainability and well-being become linked.

However, secondly, an adaptation based on the idea of well-being cannot succeed without economic growth. It is dubious that we can "manage without growth" (Victor 2008; see Jackson 2009), because tackling climate change and adapting to it is *costly*. Rich societies are better equipped to assimilate its impact than poor ones. As Nordhaus and Shellenberg note, environmentalism has always seen the economy as the *cause* rather than as the *solution* to ecological problems (Nordhaus and Shellenberg 2006). But, as a historic perspective shows, we can only be green while being rich. Neither the current understanding of economic growth nor the measurement of GDP for that reason should be exempt of criticism or amendment – changes can and ought to be made in order to reflect the environmental cost of economic activities. Yet the temptation to design people's well-being in a particular or detailed way should be avoided. It is rather a set of *objective* conditions of living under which *subjective* life-plans can be individually pursued that should be linked to climate change adaptation and hence to sustainability. For those conditions, which can be generally equated with the standards of current advanced societies, to be met, economic growth will remain necessary and desirable.

Also because, thirdly, the idea that some sort of steady-state economy can be achieved and maintained is just a delusion. Sustainability must mirror the human condition: a dynamic type of development that by its very nature is open to further transformation (see Becker et al. 1999: 6; Gallopín and Raskin 2002: 6). Although technological change and economic development can be orientated towards sustainability, it is wishful thinking to believe that they can just be stopped by decree. Governments must design markets and create the institutional conditions that eventually lead to a reasonable mitigation and to a successful adaptation, but they should do so without pre-determining a particular direction, although at the same time they must make sure that certain minimum targets are met (see Patt et al. 2010). It is all a matter of creating an institutional and economic inertia that pushes business and citizens in the direction of sustainability. To some extent, we live now in a transitional time. In fact, notwithstanding the key importance the institutional and economic drivers, it is probably the gradual cultural change induced by the current global debate on global warming that will accelerate the transition to a greener, yet liberal and open, society.

In sum, the kind of approach that climate change demands coincides with the foundations of an open view of sustainability. That is why reframing environmentalism entails reframing climate change: freeing it from the rhetoric of doom and incorporating it into a narrative of social refinement. Certainly, saying that climate change should be seen as an opportunity instead of a threat sounds like a cliché. But it happens to be true – or, to be more accurate, it can be made true.

PART III
Towards a Green Liberal Society

Chapter 6
Green Politics, Democracy, and Liberalism

Towards a Convergence of Liberalism and Environmentalism

Although we live in liberal societies, the lack of faith in liberalism is noticeable. Yet it is maybe *because* we live in liberal societies that nobody seems to be satisfied with the latter, nor with the liberal democracy that constitutes its institutional expression. And it is hardly surprising that environmentalism itself counts, from its inception, as one of the fiercest critics of liberalism – blamed for being as much a cause of the ecological crisis as an obstacle to its resolution. Such disagreement is not only practical, but normative too: environmentalists tend to refuse both the epistemological and philosophical foundations of liberalism. The result is predictable: "The majority of green philosophers and political academics remain convinced that a sustainable society would necessarily be closer to a socialist society than to a liberal, capitalist society" (Porritt 2005: 67). Such incompatibility has thus been taken for granted, all the more since environmentalism has often sought inspiration or been influenced by a number of political doctrines, from socialism to anarchism and feminism, that also stand at odds with liberalism. As a result, green political theory has been largely forged *against* liberal political philosophy, whereas, by the same token, green models of democracy have sought to overcome or at least transcend liberal democracy.

Yet something has changed in this regard during the last decade. The green commitment to democratic procedures, together with a growing, albeit minority, dissatisfaction with classical environmentalism's naturalistic stance, have brought about some reflection on a new sort of challenge: how to carry out a green programme in a liberal framework; a framework which, in turn, does not seem so hostile to green goals anymore. Green parties exert their influence in some European countries, while at the same time liberal society has made room for some sustainability claims: international agreements, technological innovations, spreading of the green lifestyle. Not a structural shift at all, to be sure, but a greener liberal society has started to look like a real possibility.

The ensuing debate was initially dominated by the conceptual revision of a number of liberal institutions – representation, rights, autonomy and citizenship – that were assimilated by environmentalism in order to achieve a greener liberal democracy. It was intended as a change from within liberalism. For mainstream environmentalism, though, this operation was not far-reaching enough. Liberal democracy had to be replaced by a more radical, genuine green democracy. Otherwise, neither ecocentrical ethics nor the right version of sustainability, i.e. a strong one, would ever be implemented. The reasons why liberal democracies are unsustainable should be found in the very foundations of liberalism (see De Geus 1999). It therefore follows that sustainability can only be achieved by removing and substituting them. If that is the case, the normative disagreements between liberalism and environmentalism would be insurmountable (see Vincent 1998: 456), and sustainability could not be liberal, nor could liberalism be green. Being green, after all, is, or should be, more than being just liberal.

But this debate has taken an unexpected direction. It now includes the opposite possibility that a liberal society becomes sustainable without embracing the core substantive values promoted by environmentalism – to be summed up as the moralisation of the socionatural relationship. In turn, such a successful green liberalism could make environmentalism look somewhat superfluous, insofar as being liberal could suffice in order to be green. Is this a reasonable expectation, though, or just an exaggeration?

In view of the radical programme of classical environmentalism, I would like to suggest that a convergence between environmentalism and liberalism can only take place between a renewed green politics and a liberalism that is structurally open to a gradual reformation of society. More exactly, the constitution of a green society – in itself a process rather than a fixed result – should be based on the free search for sustainability within the liberal-democratic framework. To support such conclusion, an inquiry is to be made into the relation between sustainability, liberalism and environmentalism. In other words, the possibility of either a liberal sustainability (Mills 2001) or a green liberalism (Wissenburg 1998; Hailwood 2004), as well as the wider question of their theoretical convergence (Barry and Wissenburg 2001; Levy and Wissenburg 2004a), is to be explored. The question is whether sustainability, understood as a conception of the good, is compatible with the liberal principle of neutrality, which tries to guarantee that individuals are free to choose and develop their own life-plans.

A Note on Liberal Democracy

But a brief note on liberalism and liberal democracy first. It is interesting to note that the latter is usually criticised for lacking democracy and containing too much liberalism (see McPherson 1964; Barber 2004). That would explain its alleged inability to solve a problem like the ecological crisis, which requires precisely the opposite. Yet it seems to me that this common position reveals a biased understanding of both the theory of political liberalism and the practice of liberal democracy.

Liberal democracy is certainly liberal rather than democratic. It is, to be more exact, a reinvention of the classical ideal of democracy that restricts the direct participation of citizens in the making of institutional decisions. It does so out of principle and out of practical reasons. Norberto Bobbio has suggested that modern democracy is indeed an extension of liberalism, provided that we understand democracy in procedural rather than substantial terms: more a government *by* the people than a government *for* the people, equitable opportunities rather than economic redistribution (Bobbio 1990: 31–7). Thus democracy is an important component of a liberal democracy, but only as relevant as some others, be they the rule of law or the separation of powers. Such political arrangement stems from a number of philosophical, even anthropological, assumptions.

According to the liberal tradition, a just social order rests on the primacy of individual freedom, the greatest threat to which is state coercion. Such freedom is an inalienable, albeit not boundless, individual right. Yet liberalism is not so much interested in proposing a *transcendent* explanation of rights, irrespective of their historical and social content, as in underlining the key importance of the pre-state, even para-state, realm. From Locke to Hayek, the *anteriority* of individual rights is suggested as the very heart of political liberalism, in a double sense: normative and practical. Needless to say, such anteriority is not to be taken literally, as a metaphysical or historical fact, but as a metaphor signalling that rights belong to individuals and cannot be arbitrarily violated by the state.

From a normative standpoint, then, such anteriority means that basic rights are not a state's property, but the very reason why the state exists: the main role of governments is to protect people's rights (see Boaz 1998: 127). But it also entails a limit to the exercise of power, because an *artificial* law – the state's legislation – should not contest the *natural* law that is already present in the customary law that precedes the constitution of the state and to a large extent coexists with it afterwards. Not everything that we do is subjected to the law. It is noticeable that the metaphorical machine of political liberalism works here at full capacity: although the state has to protect a number of rights that could not actually be enjoyed without such protection, that very state is denied any recognition as the original source of those rights. This balancing act has much to do with the ambiguous liberal distrust of human nature. This is a peculiar distrust indeed. Liberal moral expectations vary depending on the layer of social life that is in each case involved: distrust of human beings at the individual level, especially in connection to the enjoyment of power; scepticism about the existence of absolute truths or forms of life that are better than others; but optimism about the general progress of societies. All of which is ultimately plausible because if we examine reality, the prognosis seems to be confirmed: *eppur si muove*. And this leads to the practical aspect of the social's anteriority.

As it happens, political liberalism claims that it is not only *fairer*, but also more *effective*, that both individuals and societies are organised according to a general principle of restricted state intervention. A society based on central planning is doomed to fail in delivering goods and services, not to mention the lack of legitimacy that afflicts any attempt to direct their happiness. It is here that a liberal key concept, that of the spontaneous order, appears. Now, we are too accustomed to see the market – from Smith to Hayek and Von Mises – as the more eminent instance of such kind of order, but civil society, intertwined with the market and the state, is another. In both cases, the general idea is that the resources are optimally distributed without a rational planning that is unable to adapt to the myriad changes that occur beyond the reach of the most capable bureaucracy. It is noticeable that this view of society is far from the Hobbesian pessimism: competition co-exists with cooperation, self-interest with altruism. Or so it is claimed.

From this point of view, democracy is not conceived of as a tool for the citizen's participation in collective decision-making, but as a means to limit state power on behalf of individual freedom. Political liberalism is not so much a general ethics, as a procedural framework within which individuals can pursue their conception of the good. Not a neutral framework, certainly, as it is full of substantive values. We will return to this later. But the goal of political liberalism is to create a political framework that avoids the concentration of power and thus liberates as much space as possible for the spontaneous operation of the social order. A general preference for society against the state is ostensible. And such preference is due to the primacy of the individual against those abstract supra-individual entities that so often have been levelled against him: social class, nation, race – and nature as well?

Rights and liberties, separation of powers, checks and balances between them, political representation, a separation between the public and the private sphere, public neutrality regarding the different conceptions of the good: such are the essential principles of political liberalism. They are constantly invoked as the very centre of the liberal tradition, contributing to its epistemological coherence. Yet this should not be identified with rigidity on the part of political liberalism. There are a number of liberal traditions, from utilitarianism to libertarianism and social liberalism, that convey different interpretations of those principles and of the way they might be organised: De Jasay and Rawls are very far apart. Besides, liberal democracy has shown a great ability to assimilate new ideas and claims, a trait that has decisively contributed to its survival. It also helps to explain the transformations it has experienced.

In this regard, it is the morphological evolution experienced by the liberal state during the first part of the twentieth century, which turned it into a welfare state, which must be underlined. Although Western political systems are still mostly based on the principles of political liberalism, liberal democracy has not only made room for a greater range of citizen *demands* before the state, resulting in a number of new entitlements, but also for a notable enlargement of the *form* through which those demands are carried out. The claim that citizens participate in the shaping of their society just voting periodically reveals a narrow understanding of society's *modus operandi*. To begin with, both citizens and their associations – be they civic, political or economic – are constantly *acting* within society, thus giving shape to it spontaneously. This is a private, not planned performance, with public consequences. But it also entails a private deliberation that produces new ideas and values: it shapes a society that does not adopt a definite shape. In fact, on the other hand, liberal society has gradually opened itself to new forms of political participation, both formal and informal. It is the case with local participation, the normalisation of collective action, or the war of meanings that take place daily in the realm of mass communication. They are new ways of expressing substantive and symbolic demands – *from* society *to* the political system.

None of this entails a blind endorsement of current liberal democracies. It is obvious that they are not perfect – how could they be? This is especially true concerning the *realities* of social pluralism, as opposed to a *normative* pluralistic ideal that cannot be said to be fully realised. Political decision-making is constrained by economic or institutional actors and inertias that actually limit the range of conceivable solutions to social problems. Yet this reasonable objection does not diminish the case for liberal democracy, but rather it reinforces the case for reforming it. In this regard, the communicative revolution that the Internet has brought about is starting to coalesce in new, more collaborative and less hierarchical ways of doing all kinds of things, from business to political protest and government, whose consequences for the institutional shape of democracy remain to be seen. Thus there is no reason to believe that the current shape of democracy is not going to be altered: liberal democracies have evolved. It is not worth discussing now whether such gradual transformation has occurred *despite* political liberalism or *thanks* to it. The fact remains that such adaptability on the part of liberal democracy should moderate the calls to overcoming it, including the green one. In fact, this flexibility can be used for reconciling environmentalism and liberalism. Green politics can be more liberal without ceasing to be green, thus making possible, precisely, a green liberal society.

Liberal Neutrality and Environmentalism

The disagreement between political liberalism and environmentalism revolves largely around the relationship between the liberal principle of neutrality and the green conception of the good. Again, a conflict arises whenever a procedural framework and consequentialist ethics meet. It goes without saying that the conflict concerns sustainability too. Searching for sustainability is, after all, assuming publicly a substantive conception of the good – the green one. Or is it not?

Naturally, the more openly sustainability is defined, the more easily a solution to this problem may be found. If, on the contrary, a strong version of sustainability is promoted as *the* public version of the principle, such convergence looks non-viable. It might be the case that it is liberalism that must adopt a moral view of nature if it is to become sustainable. The reason is simple: liberalism is founded on an anthropocentrical epistemological framework that hinders the adoption of greener – i.e. more ecocentric – attitudes on the part of the citizens. Robyn Eckersley suggests that "the liberal state reinforces a particular kind of self, with particular kinds of dispositions" (Eckersley 2004: 105). In this case, such disposition is no other than seeing nature as a resource for the satisfaction of human needs, either material or aesthetic. And that is why environmentalism wishes for a different context, where better – greener – citizens can emerge, leading to tighter environmental standards and better – greener – policies: a virtuous circle. Thus:

In addition to a concern with acting *effectively* within a given political institutional context, environmentalism is also engaged in redefining and *reshaping* that context. And in addition to its concern with *institutional* design, environmentalism is also engaged in specifying and defining the environmental *goals* those institutions should promote, goals like the preservation of a self-sustaining nature or natural biodiversity (Levy and Wissenburg 2004b: 4).

Yet these goals require not only procedural, but also substantive means. Whereas liberalism promotes axiological neutrality and proceeds to aggregate autonomous individual preferences, environmentalism demands some kind of state intervention, on account of the need to protect a common good, i.e. the environment. Thus it defends a particular conception of the good. Mark Sagoff put it as follows:

If the laws and policies supported by the environmental lobby are not neutral among ethical, aesthetic, and religious ideals but express a moral conception of people's appropriate relation to nature, can environmentalists be liberals? May liberals support environmental laws even when these conflict with the utilitarian and egalitarian goals we usually associate with liberalism? (Sagoff 1988: 150).

More than 20 years later, the problem has not changed much (see Humphrey 2004: 125). This certainly goes to show that it is not easy to solve – namely, that maybe liberalism and environmentalism are irreconcilable. Sagoff's own answer is a good guide to the issue's intricacies. He sees two basic distinctions in liberalism, namely, a separation between the state and the civil society on the one hand, and a separation between the basic institutional structure and the particular social policies that result from it on the other. Liberals claim that the institutional structure of society must be neutral regarding the different conceptions of the good held by their citizens. Thus liberal theory, in itself a comprehensive moral view, applies on that level, but not on the level of the resulting social policies. The institutional structure regulates the debate and negotiation *between* different conceptions of the good. Henceforth, environmentalists can and must be liberals. As liberal institutions do not anticipate any particular political outcome, environmentalism can fight to produce them in the layer corresponding to social policies, where the rule of neutrality does *not* apply. Therefore, Sagoff concludes, liberal democracy turns out to be the most favourable form of democracy for achieving green goals.

From this point of view, liberalism can be said to articulate the political system by distinguishing means from ends: the former operate neutrally regarding the latter. Therefore, the competition among conceptions of the good can only be expressed via particular social policies, not via a transformation of the general political framework – lest neutrality is violated. The environmentalist can promote green policies, but should not try to impose a general moralisation of socionatural relations. Following Joseph Raz (1988), a *neutrality of justification* would demand that political procedures and actions are not justified as the product of a given conception of the good, whereas a *neutrality of outcome* would prevent that the political process results in the promotion of a conception of the good over others. The former is acceptable; the latter would just lead to social paralysis.

But then a problem arises. Environmentalism *is* a consequentialist conception of the good, one that has traditionally aspired to a thorough transformation of society. Such purpose is embodied in the strong versions of sustainability. Some of them even ban travelling! Green goals thus seem to go beyond the liberal regulative framework, which in turn means that the mere competition for social policies might not be enough for them. Is this not an insurmountable obstacle for any agreement between environmentalism and liberalism? Going against Sagoff, then, greens *could not* be liberals. Some kind of green reformism might have a place within the liberal framework, but not the robust conception of the good that environmentalism actually is (Dobson 1998: 12). Indeed. But that does not refute political liberalism as much as it shows the need to renew environmentalism, in order to make their convergence feasible. Admittedly, classical environmentalism would have it the other way around, claiming that it is liberalism that needs to be amended. But let us see.

The neutrality principle stems from the liberal conviction that there is not any true conception of the good. In other words, liberalism does not support any epistemological utopia, believes not in a closure of knowledge (see Kolakowski 2006). As it is impossible to choose any single conception of the good among many, a neutral political structure is set up that guarantees those basic freedoms that permit individuals the satisfaction of their preferences, derived in turn from their own conceptions of the good, in the private sphere. The state cannot influence individual preferences: "People's needs – for company, children, food, technology, travel and trinkets – are private affairs; control, if possible at all, is impermissible" (Wissenburg 1998: 67). However, environmentalists question the very separation between the public and the private upon which liberal neutrality rests. Taking care of the environment would require the possibility of *interfering* with private choices (see Smith 1999: 52). Individual autonomy can easily clash with the requirements of sustainability. And vice versa: sustainability can be a threat to liberal neutrality, if it demands a restriction of liberties incompatible with the liberal order.

This problem has been exposed by recurring to the somewhat simplistic distinction between citizens and consumers. If environmental goods are public goods, they cannot be left to the dynamics of the market – they must be protected. There is supposed to be a difference in the formation of individual preferences towards public and private goods. By deciding about public goods, we must weigh our interests as much as the others, taking into consideration our ethical principles and the kind of society we prefer (Jacobs 1997: 219; Elster 1997). As consumers, we just think of our own benefit; as citizens, our decision is grounded on the public interests of the community. That is how things *should* be. And such is the implicit premise in the green view of the good:

> A green life is not primarily a better life (although it might be that too). The value of living responsibly with regard to other beings and things resides not so much in what it allows the individual to be as in what it allows the world and our society to be (Beckman 2001: 184).

Sustainable attitudes would not reflect as much preferences as ideas of the good. Therefore, environmental policies should not be based on the preferences as expressed in the market, but on the values that emerge from the debates about the public good (De-Shalit 2000: 90). Again, so it should be!

But environmentalism finds yet another problem in the liberal society, namely, that environmental goods are not *perceived* as public goods. And such perception is taken as a necessary condition for the formation of community-orientated attitudes. If values and preferences do not reflect the particular individual, then they reflect the society in which that individual lives, so that the society in which we live matters enormously. This leads to the well-known green argument, rooted in Marxian epistemology, according to which the liberal-capitalist context hinders the emergence of green values – or deep green values, to be precise. Furthermore, that implicit bias would be worsened by the liberal indifference towards the preference formation process. The green retort is obvious: neutrality ends up reinforcing the *private* perception and treatment of *public* goods, thus becoming a great obstacle to the achievement of sustainability. That is why environmentalists criticise the social and communicative context in which individual and social preferences are formed and enacted (Eckersley 2004: 96). Moreover, "the quest for sustainable modes of being represents a challenge to liberal democratic notions of how the collective good is determined" (Davidson 2000: 34). If we want the individual to behave himself more as a citizen than as a consumer, it seems, we have to *educate* him.

Nevertheless, the distinction between the roles of citizen and consumer is a bit too lazy. Is it that clear that we, as citizens, are committed to the public interest and to our conception of the good, whereas we, as consumers, remain exclusively attached to our private satisfaction, mysteriously rendered as something separated from our conception of the good? Such categorical separation can actually happen only in an *ideal* individual – an individual that, by definition, does not exist. There is no such thing as consumption devoid of social and symbolic elements beyond the simple cost-benefit analysis. Of course, *taste* is a different matter, but the lack or the abundance of taste is ultimately dependant on education or refinement. Moreover, this is a trait that will be evident in *any* realm of an individual's life, not only in their environmental attitudes. To believe that our subjectivity possesses closed compartments is self-deluding. There is no public reasoning without private insights, and vice versa.

Moreover, if we accept the suggestion that a given social context is *decisive* for determining individual preferences, the very value pluralism upon which the neutrality principle rests would be dramatically undermined – because there would never be such thing as an individual *choice* anymore. Yet which preferences and values are chosen, which are exogenously determined? If the liberal-capitalist context induces a particular type of them, is the same not bound to occur in any other social context? Could it not even be the case that some other social contexts (like a strong sustainable society, to name but one) would restrict the range of conceivable preferences more tightly? After all, it is paternalistic to suggest that citizens are not aware of their own preferences, especially since there would be no objection at all were such preferences to coincide with those that their critics hold. As Mathew Humphrey has pointed out, "Because *I* completely fail to comprehend the notion of a life devoted to conspicuous consumption, does this give me a reason to somehow rule it out of court as an acceptable life-plan?" (Humphrey 2002: 59). A preference for nature's protection and enjoyment should then be deemed "as a respectable personal ideal – but nothing more than that" (Birnbacher 2006: 131). However, the same does not apply to sustainability writ large, since the latter is, ultimately, a precondition for the development of *any* life-plan.

Henceforth, in the absence of a perfectly neutral social context, political liberalism aims to solve the conflict between different conceptions of the good by setting up the principle of neutrality in the institutional level. Yet that does not mean complete political neutrality on the part of the state: democratic societies possess substantive rules that are the outcome of a gradual assimilation of socio-political debates on the good life. There is a safety net for the poor and unemployed; there are rights for the protection of minorities; laws exist against animal abuse; and so on. As Sagoff claims, "The satisfaction of preferences per se (...) is hardly the principal purpose or policy goal of a civilised society" (Sagoff 2004: 6). True. And what lies outside that satisfaction is precisely what is incorporated into the general principles of that society. As Rawls (1993) claims, there is a moral dimension in the political community. And the interventionist liberalism that is dominant in the *practice* of Western democracies articulates that dimension. Liberal political institutions are hence not impassive before social change. They admit exceptions to the principle of neutrality. A democratic society is thus an ethical-political structure some of which ongoing debates produce changes in the institutional level. But these changes are not exclusively decided within institutions, nor are necessarily translated into legislation. Sustainability might just be one of those exceptions, as long as it is understood normatively – as an open sustainability.

Sustainability and the Rhetoric of the Good Life

The green critique, certainly shared with other political doctrines that aim for a substantive transformation of society, exposes the flaws implicit in the very idea of neutrality. They can be summarised by saying that proposing neutrality is not the same as achieving it. Liberalism is in itself a comprehensive moral view that does not permit the full realisation of any other rival moral view. It is a sort of aporia: liberalism may be said to be a conception of the good grounded on the impossibility to choose between different conceptions of the good. Yet it is in itself biased in favour of these values – liberal values – which are what makes the discussion on the good life possible at all. And although it does not try to impose an homogeneous morality, the neutrality rule promotes *de facto* a particular conception of freedom *as* the good life, thus generating a dynamic that favours some moral developments above others (Stephens 2001: 7). Therefore, liberalism is not so much a procedural principle that regulates the public dispute between rival conceptions of the good life, as a conception of the good in itself.

A question follows. To what extent does this rule restrict the realisation of those conceptions of the good that, like environmentalism itself, more strongly disagree with the current social order? Marcel Wissenburg (1998: 61) has pointed out that liberalism does not seem authorised to prescribe a single sustainable society, since that would mean choosing a conception of the good among many. Instead, it can only conceive the sustainable society as a range of possible and permissible worlds, not as a sacred goal of human existence. This is in turn connected to liberal scepticism. If there is no definite truth, the dispute between different conceptions of the good cannot conclude with the full realisation of any of them. As a consequence, the outcome of the debate seems somewhat irrelevant. Alasdair MacIntyre has thus described the liberal order as:

> one in which each standpoint may make its claims but can do no more within the framework of the public order, since no overall theory of the human good is to be regarded as justified. Hence at this level debate is necessarily barren; rival appeals to accounts of the human good or of justice necessarily assume *a rethorical form* such that it is as assertion and counterassertion, rather than as argument and counterargument, that rival standpoints confront one another (MacIntyre 1988: 343, my emphasis).

Substantive conclusions would then lack any real importance; the debate would be condemned to circularity. In fact, it is this atmospheric impartiality that would prevent people from thinking in terms of the "good life" (see Dobson 1998: 207). It is paradoxical, though, that the same foundations of liberalism that have allowed environmentalism to thrive would now be signalling its limits. Henceforth, although liberalism's anti-chauvinism has made the promotion of green attitudes and values possible, it also sets a limit for the public realisation of the green conception of the good (see De-Shalit 2000: 66). That is why environmentalists denounce that the ensuing liberal neutralisation – via assimilation – of the green programme represents but the victory of environmental pragmatism over ecocentric theory (Eckersley 2002: 49). But is the liberal rhetoric really a trap?

Liberalism deems acceptable those partial outcomes of the public debate that do not threat an institutional structure grounded on the premise of impartiality towards the different comprehensive conceptions of the good. The reason is that the full adoption of any of them would *prevent* the continuation of such debate. It can be put this way: the neutrality principle creates the conditions for the public debate, but at the same time it postpones indefinitely the chance that totalising outcomes can be ever derived from it. The consequences for the green agenda would be harmful:

> The paradox, then, is that, while on the one hand liberalism allows and encourages discussion on environmental issues, on the other it cannot permit the outcome of the discussion namely the implementation, maintenance, and justification of environmental policies. Thus, it precludes constructive public action that is meant to protect the environment (De-Shalit 2000: 65).

Yet is that so? Although they are always bound to be considered unsatisfactory by greens, there is no shortage of environmental policies within liberal democracies. However, if that were the case, is it necessary to reformulate the principle of neutrality, so that the latter allows for adopting and implementing substantive conclusions, provided that such adoption does *not* suppress the chances of other conceptions of the good to be adopted as well? Such is Seyla Benhabib's suggestion: the adoption of a constrained neutrality consisting of rules which are so abstract that different ways of life and conceptions of the good are allowed to flourish, while at the same time none of them can behave in an authoritarian manner regarding others (see Benhabib 1992: 16; Downing and Thigpen 1989). It is not clear, though, how this could work.

It is not, because conflicts between rival conceptions of the good are more common than agreements. If a conception of the good demands social substantive changes – as a strong sustainability would do – the conflict is unavoidable and actually irresolvable. By admitting exceptions to the general rule of neutrality, liberalism already guarantees a balance between fairness and goodness. Political debates and negotiations translate to an institutional level the larger debate that takes place in society. They do so in a twofold way: as an argumentative debate in the realm of communication and as a performative debate in the realm of action. The fact that green values have flourished in the last decades proves that liberal democracy does not prevent socio-cultural dynamism.

On the other hand, green commentators have demanded a particular type of liberalism for the dialogue with environmentalism to bear fruits at all, i.e. a liberalism rooted in the social-pragmatic tradition (see Musschenga 1994; Wissenburg 1998: 74; De Geus 1999: 35; De-Shalit 2000: 92; Stephens 2001: 2; Barry 2001: 79). In other words, a pragmatic liberalism that sees democracy as something more than a tool for *protecting* citizens – namely, as an ethical community also devoted to foster the moral and personal *development* of those citizens. Instead of a purely formal liberalism that sets up a sharp distinction between politics and society, a substantive liberalism would permit state interference whenever it is justified (Levine 1981: 23). Public debate should then go beyond mere rhetoric, although the following interference cannot unjustifiably constrain the individual search for the good life. Such is, in sum, the view of liberalism that environmentalists have got used to invoke: a liberalism that fosters public debate on the good life and is therefore ready to make substantive institutional changes.

However, is this an interpretation of the *desirable* liberalism, or an approximate description of the way in which liberal democracies *actually* operate? Because, for all our theoretical assumptions, the principle of neutrality does not hinder a substantive political debate, nor therefore are there insurmountable obstacles for achieving sustainability. Society changes, an imperfect debate goes on, sustainability is discussed. This may be a slow, disappointing process – but it is the slow, disappointing pace of democracy. Yet, on the other hand, environmentalists may also be frustrated because there seems to be no place for their preferred, strong versions of sustainability within liberal society. This is true. Some versions of sustainability are not compatible with the principle of neutrality and are even openly illiberal. And between liberal neutrality and radical sustainability – between the fair and the good – the former must prevail.

Open Sustainability and Green Liberalism

Nevertheless, not even an interpretation of neutrality that allows a certain degree of public interventionism *guarantees* that green values will prevail as a result of the debate on the good life. Should sustainability then be incorporated to the core of basic democratic values, on the grounds that it is a precondition for liberal society itself?

This has been suggested. The rationale for this option is clear: liberal democracy can make sure that the socionatural relationship is sustainable without contradicting its commitment to neutrality. After all, some interests are already protected within liberal democracies, so that sustainability could also be justified as a necessary contribution to the preservation of the *liberal* conditions of society. In this vein, nature's protection would be justified as a protection of neutrality itself, insofar as the right to enjoy a sufficiently "natural" environment should be protected (Humphrey 2002: 189; Birnbacher 2006: 190). This caution is valid not only for present, but also for future people, since we cannot know whether they will wish to enjoy nature or not (Ott and Döring 2004: 104). The preservation of a non-anthropogenic nature would then serve the purpose of protecting the *possibility* of a green way of life and hence the diversity of the moral landscape. It is relatively unimportant here whether nature is protected on ecocentric or anthropocentric grounds: a sound environment would have a *derivative* value for liberal society (Vincent 1998: 447–8). From this standpoint, liberalism should not neglect green demands, lest it betrays its own foundations. A liberalism without sustainability would then not be conceivable. Yet a sustainability without liberalism *is* conceivable. And that, precisely, is the problem.

Once again, the point is that nature's protection is not the same as sustainability. The former can embody different degrees of the latter. So which one is exactly going to make its way into the core principles of liberal society? The implications differ. It should be remembered that the justification for greening liberal democracy is directly connected to the *form* that sustainability is to adopt. And different sustainable societies carry different consequences for the shape of society and the lives of its citizens. Or, to put it differently: we may well say that nature's protection or sustainability should be incorporated into liberal democracy's core values, but which degree of protection and which version of sustainability?

After all, sustainability *as such* does not require a special justification anymore. Almost everybody is inclined to agree on the need to re-organise the socionatural relationship in a durable way, a conviction that the threat of climate change has reinforced. Of course, particular justifications can be provided. For instance, Derek Bell (2005: 183–5) has suggested that the protection of both basic needs and reasonable pluralism demand a certain degree of environmental protection, whereas Simon Hailwood (2005: 199) claims that contemporary political reasonableness involves an environmental commitment that does not undermine liberal neutrality. In this way, we state that liberalism must be sustainable and a minimum degree of naturalness should be protected. Yet the precise *content* of sustainability remains undetermined. It is a greater protection of nature, or even more clearly, it is the implementation of strong versions of sustainability that demand a greater justification. In fact, each particular version of sustainability should be justified before the equal pretension of rival versions.

Henceforth, the solution lies in adding to the institutional structure of liberal democracy a normatively conceived principle of sustainability – that is, a sustainability understood as an open principle towards which society is orientated, without predetermining its particular content. As it happens, a closed interpretation of sustainability is not compatible with the open and procedural character of liberal democracy, since it violates the neutrality principle and would prevent the democratic emergence of the sustainable society. In turn, this means that strong versions of sustainability can be defended *within* liberal democracy, but cannot be fully adopted if that means the closing of that very debate. Whereas although a radical environmentalist can make his voice heard inside liberalism, nobody's reasons would be heard in a radical sustainable society.

On the contrary, once the intrinsically normative character of sustainability is recognised, sustainability and democracy can be seen as mutually reinforcing principles. Thus liberal democracy provides the neutral framework wherein the debate about the shape of sustainability takes place. Sustainability becomes a core democratic value that is implicit in the foundations of liberal society. The former's normative quality does not determine *ex ante* the shape of the (liberal) sustainable society, rather it makes this shape dependent on the conclusions – always provisional – that derive from such debate. Society is orientated to sustainability, but not forced to adopt a particular version of it.

Admittedly, the idea that the liberal state *will* eventually adopt sustainability as a general goal may seem delusionary, since there is no guarantee that this will actually happen. In fact, the current degree of consensus on the need to pursue sustainable policies cannot be exaggerated, especially in the aftermath of a financial crisis which has brought a considerable increase of the unemployment rate. However, environmental concern is growing in every realm and it seems a matter of time that this general trend becomes more or less formally entrenched in the constitutional *ethos* of liberal democracies. A general mandate towards sustainability would meet resistance, of course, because it would impose some minimal ecological standards that would operate as the framework within which particular versions of sustainability would contend.

An improbable parallel can be drawn in this point. Remember Robert Nozick's framework for utopia? I suggest that it can help us to devise the kind of liberal sustainable society that we are trying to sketch. Nozick's interest was the maximisation of freedom in the pursuit of life-plans with minimum state coercion. He went on to claim that a minimal state should be the most appropriate institutional organisation for fulfilling that end, since its main function would be the policing of voluntary contracts that different individuals and their associations would sign in order to realise each owns conception of the good. Hence: "There is the framework for utopia, and there are the particular communities within the framework" (Nozick 2008: 332). Insofar as basic rights are not violated, nor the general institutional structure neutralised, there would be no limits for the coexistence of disparate life-plans. Of course, this framework is partly a utopia in itself, but it also reflects the underlying structure of liberal societies. Yet the most interesting insight for our purposes is the way in which it combines the general, overriding primacy of freedom as an abstract principle with the emergence of communities within which that very freedom could be partially restricted with the consent of their members: e.g. a Maoist community would be authorised to live within a liberal (or libertarian) society.

Now, a similar scheme can be applied to our topic. Inside the green liberal society, the argumentative debate about the most desirable version of sustainability would run parallel to the individual and collective action related to the practical achievement of sustainability: sustainable lifestyles, scientific innovation, international agreements, corporate adaptation, environmental education, and so on. We may see each of these sustainability-related actions and ideas, enacted or defended in both the public and private spheres, as embodying a particular version of sustainability. They are generally orientated to sustainability and particularly orientated to a specific version of it. In other words, there is the framework for sustainability, and there are the particular sustainabilities pursued within the framework. Whereas the general primacy of freedom is guaranteed in Nozick's libertarian society by the minimal state, the liberal state in a green society must make sure that the latter's general orientation to sustainability is minimally successful in every given moment, that is, must demand that some requirements – for instance in matters of conservation, pollution, substitution – are met, while at the same time individuals and associations are free to practice and promote their preferred conception of the sustainable society. It is obvious then that the comparison cannot be drawn until its completion, because sustainability requires a greater coordination in view of its structural challenges. But then again, no minimal state is vindicated here, so the comparison is, I would like to suggest, useful nonetheless. So:

Society is generally orientated towards sustainability, but no pre-given model of the latter is to be enforced, rather the sustainable society will be the result of the ongoing debate and action that take place in both the public and the private spheres with the supervision of the state, so that citizens and associations are free to pursue their own conception of sustainability, provided that, by doing so, they do not hinder the equal freedom of others to do the same, nor violate the public mandate to meet some minimum environmental standards.

Thus citizens are not forced to be green, but are encouraged to behave responsibly in a social frame where sustainability is institutionally recognised as a societal goal. However, the lack of a given blueprint for sustainability that is generally adopted means that citizens are both free *from* sustainability and free *for* sustainability: they do not have to comply with a particular conception of sustainability, although they must obey the environmental laws gradually adopted by the state, while at the same time they can pursue and promote their chosen view of a sustainable society.

Therefore, the desirable convergence of liberalism and environmentalism depends on the adoption of an open conception of sustainability. Thus the relationship between liberalism and green politics adopts a symbiotic character: liberal democracy provides the neutral framework wherein the principle of sustainability, incorporated into the values that guarantee the continuity of the former, is discussed. Henceforth, a viable socionatural relationship is adopted as a general orientation of society. Sustainability, openly understood, can then be reformulated as a liberal principle. Thus:

As to the *normative foundations* of both liberal democracy and environmentalism, there are good grounds to believe that environmental and liberal democratic values can cohere within a pluralistic conception of a reasonable, just and green society (Levy and Wissenburg 2004c: 195).

Although liberalism may not foster green values and lifestyles, it does not hinder their development either. After all, the former have emerged *but* in liberal contexts. And it is plain that liberal societies – thanks to or in spite of liberalism itself – are growing more concerned with the environment and sustainability. The compatibility of a renewed environmentalism and liberalism is not just a theoretical possibility, but an incipient reality.

Chapter 7
Can We Democratise Sustainability?

The Greening of Liberal Democracy

So, what is a green liberal society like? Interestingly, the only possible answer to such question is that we cannot know. The reason is that a green liberal society is just like our current Western societies, plus an explicit recognition of sustainability as a public principle. The shape that the debate and practice on the latter will lend to society cannot be known in advance, although we can of course make some inferences from current societal and technological trends. Of course, this premise is deeply unsatisfactory for environmentalists, because it seems to leave things as they stand, blocking any change in the direction of a more ecocentric society. Yet it does not have to be so. So long as the basic principles of a liberal society are not violated, society can end up being as green as it wishes. It just remains to be seen.

However, this is not to state that in order to make the achievement of sustainability feasible liberal democracy cannot be amended at all. In fact, the recognition of sustainability as a public principle may demand some institutional changes, lest its achievement is frustrated, while at the same time the open nature of the principle limits the reach of such changes. Again, this will frustrate environmentalists, because they tend to think that an institutionally promoted change is more reliable than the one generated within society, especially if the latter is capitalist and the preferences of their citizens are mediated by the market. This is why green political theorists have persistently attempted to absorb and redesign liberal concepts and institutions: to make liberal democracy greener and thus advance the cause of a strong sustainable society.

Although this theoretical operation has taken different forms and affected a number of different liberal institutions, its general meaning is clear: to build up a green democracy that goes beyond, or that at least deeply corrects, liberal democracy. John Barry (1999: 198) maintains that it would be a *post-liberal* rather than an *anti-liberal* democracy. The task is not so much reinventing, as elaborating new categories that allow us to re-interpret traditional institutions, adapting them to new demands and transformations (Vallespín 2000: 13). From this standpoint, the aim is to *extend* the reach of the classical democratic institutions, for them to cover a dimension of the social life traditionally neglected by liberalism (and most of the political doctrines so far): the relationship with nature. It is the case with political representation and rights, citizenship, the community, the state, even the market. This programme is supposed to culminate with the adoption of deliberative democracy as *the* foundation for a greener democracy.

However, the theoretical greening of liberal democracy has been too conditioned by classical environmentalism's normative ambivalence. These limitations should be left behind, in the benefit of a more reformist stance regarding a liberal political system that need not be thoroughly reinvented. It is in this way that a renewed environmentalism can contribute to the constitution of a green liberal society. The relationship between sustainability and democracy is a case in point. Naturally, the question about whether we *should* democratise sustainability is merely rhetorical, because it is obvious that we must do so *if* we wish to live in democratic and sustainable societies. Therefore, asking whether we *can* democratise sustainability becomes a more relevant question. This is especially important regarding the approach to sustainability advocated in this book. An open view of sustainability depends on the amount and dynamism of different social actors' contributions to its discussion and development. It is thus radically democratic. Yet this condition does not forcefully involve an endorsement of direct forms of democracy as the best means for achieving sustainability. Which is precisely what environmentalists normally claim.

The Promise of Deliberation

When they are questioned about the kind of democracy best suited for dealing with ecological unsustainability, the consensus within the latest green political theory seems obvious: it should be some version of deliberative democracy (Jacobs 1997; Eder 1995; Dryzek 2000, 2006; Eckersley 2000, 2002; Smith 2003; Baber and Bartlett 2005). Such a deliberative turn has paralleled that of democratic theory, so there is nothing to be surprised about – it is part of a larger picture in contemporary political theory. They converge, in fact: "The deliberative democracy movement has been spawned by a growing realisation that contemporary liberalism has lost its democratic character just as it has also sacrificed its ecological sustainability" (Baber and Bartlett 2005: 5). However, the green defence of deliberative democracy should not be limited to the goal of deepening democracy, for that is a widespread ideal for whose defence – whatever its merits are – democratic theory suffices. Therefore, either to accept or to refuse deliberative democracy should depend more on specifically green arguments on its behalf. Thus it makes sense to look for a *green* justification of deliberative democracy, rather than for a *democratic* one that can be found elsewhere outside green theory. That said, the reasons why deliberation turns out *not* to be a panacea for sustainability have to do with problems that are intrinsic to democratic deliberation itself.

In that regard, although it may be in its prime now, traces of a deliberative approach can be found throughout democratic theory's history. Deliberative democracy is not a novelty inasmuch as a development – one rooted in Greek democracy, American federalism and French constitutionalism, Stuart Mill and Dewey, Critical Theory's criticism of instrumental reason, and so forth. Contemporarily, the Habermasian theory of communicative action has contributed enormously to giving rise to a theoretical movement in which seminal contributions by Manin (1987) and Dryzek (1990) stand out, later reinforced by a number of thinkers (Elster 1997, 1998; Benhabib 1996; Bohman and Reg 1997; Gutmann and Thompson 1996). Influential contributions by Rawls (1993) and Habermas (1997) strengthened this channel of thought. Since then, the literature has experienced a continuous increase (Macedo 1999; Saward 2000; Gutmann and Thompson 2004; Estlund 2008; Gastil 2008).

Deliberative democracy is grounded on the procedural legitimacy provided by a collective decision-making method. This method is based on the equal participation of citizens in a public-orientated and bounding deliberation. Democracy is then an intersubjective dialogue among free and equal citizens, defined by their ability to engage in political participation and public debate. Henceforth, a legitimate decision is not an expression of general will, but the outcome of a collective deliberation. The possibility that individual preferences are transformed in the course of deliberation is then essential to the deliberative model, as it is the supposition that deliberation leads to the right outcomes – a twofold feature that is captured in the label "epistemic proceduralism" (Habermas 2008; Estlund 2008). How this general idea is to be institutionalised and put into practice is a complicated matter, which different approaches resolve differently (see Gastil and Levine 2005).

Be that as it may, what are the green reasons for supporting deliberative democracy, as the means through which a more sustainable society can be built upon? In short, the answer is that a deliberative democracy is expected to produce more sustainable outcomes. The reasons why can be briefly presented as follows (see Arias-Maldonado 2007).

1. *Green values will emerge more easily in a deliberative context.* This is an epistemological-pragmatic argument, according to which the open nature of a deliberative procedure, whose argumentative and rational orientation facilitates the persuasive and ordered exposition of all values and preferences, would suppress the distortions of liberal political process, thus allowing the emergence of green values – thanks to the latter's objective rational appeal. Deliberation would widen the fallible and limited standpoints of individual participants, making good use of the knowledge, experiences and abilities of the others (see Smith 2001: 73). Hence it would perform a releasing function for environmental and ethical concerns that, however strong, usually remain latent (Niemeyer 2004: 348). Thus deliberation would green decisions.

2. *Deliberative democracy incorporates excluded actors and voices into the democratic process, among them those from the non-human world.* This is the inclusive argument. Its premise is that by excluding the interests of the natural world from the political process, liberal democracy *produces* un-sustainability. On the contrary, the inclusiveness of deliberative procedures – and the very nature of the debate within them – would foster

an "enlarged thinking" that is able to incorporate such interests (Eckersley 2000: 121). Obviously, nature does not speak, so that those interests should be internalised by humans, who would in turn articulate some kind of vicarious representation of these underrepresented agents. Such representation is supposed to find a more adequate vehicle in a democracy wherein those initially distant interests are listened to and hence given proper consideration (Goodin 1996: 847). In addition to this intra-human deliberation, communicative rationality could be extended to natural entities, by treating natural signs with the same consideration agreed upon for human ones (Dryzek 1995: 21, 2000a: 149). Deliberation is supposed to green democrats.

3. *Deliberative democracy is the best institutional arrangement for developing ecological citizenship.* The reinforcement of citizenship leads to the recognition of something that early environmentalism failed to see: how sustainability and the reshaping of social-environmental relationships are political rather than moral questions (see Barry 1999: 67). The normative condition of sustainability asks for the constitution of democratic frames of deliberation and decision. A deliberative institution appears to be the most adequate political arrangement for ecological citizenship. This deliberative frame requires an active view of citizenship, in which citizens' experiences and judgments are incorporated into the public domain, and a mutual respect and understanding among them is encouraged (Smith 2000: 32). Ecological citizenship cannot then exist without deliberative politics. Deliberation greens citizens.

4. *Deliberative democracy is the best way to combine expertise judgement and citizen participation in decision-making process.* Sustainability offers two contrasting sides. On the one hand, it is undeniably normative, for it cannot be ideologically or scientifically pre-determined, but must be defined according to value judgement. On the other, it requires technical implementation through science and technology. However, the relationship between citizenship, expertise and democratic decision is hard to adjust, i.e. it is not easy to decide whose judgements must weight how much and against whom, in order to reach a final decision about what. Precisely, deliberative democracy would be well-equipped for dealing with this problem. It would allow a deliberation between citizens, politicians and scientists on environmental problems (Lidskog 2000). In other words, it would facilitate the democratisation of ecological risks without neglecting their technical dimension. Deliberation would green expertise.

5. *Deliberation leads to more legitimate and efficient decisions on sustainability.* It is to be noticed that society's complexity, which lies at the origin of social production of ecological risks, increases the contingency of any social action: the greater the possibilities for action, the greater also the interdependence of the chosen courses of action (Eder 2000: 230). In such context, a greater inclusion in the decision-making process, especially

of those affected by a specific risk, diminishes the de-legitimising effects of a decision that would finally happen to be mistaken. Legitimacy is, in this context, a function of co-responsibility. However, the introduction of deliberative devices for the definition and management of sustainability can result not only in more legitimate, but also in more efficient and rational decisions. The epistemic quality of deliberation applies to decisions on sustainability too. Deliberation greens rationality.

Therefore, in spite of being inspired by democratic theory itself, green reasons for adopting deliberative principles and procedures are neither scant nor weak. Thus the case for deliberative democracy in green politics is strong enough to be considered. It amounts to the currently hegemonic approach to democracy within environmental thinking. Yet not too many expectations should be placed on deliberative democracy as a vehicle for greening society. Deliberation should not be asked for more than it can deliver. It is not certain, in fact, that it can deliver what is supposed to.

Why Deliberative Democracy is Not a Solution

A preliminary question is to be asked: are deliberative procedures only accepted by environmentalism *as long as* they contribute to the achievement of green goals – so that any other democratic or even political model, showing more ecological efficiency, would be chosen instead if it were to be found? Although such instrumental perspective is not explicit on green accounts, it appears to underlie the otherwise diverse arguments supporting deliberative democracy. Deliberative democracy is usually taken as a means to ecological ends: a given procedure whose political virtues will eventually deliver environmental advantages. This is consistent with the temptation of consequentialism that has traditionally troubled green's relationship with democracy.

Yet this suspicion may not be fair. After all, environmentalism can only provide its commitment to democracy, not democracy's commitment to green values. Therefore, once the green commitment to democracy has been openly stated, looking for the most favourable model of democracy from a green point of view is the thing to do – and the choice is consistent with environmentalism's traditional inclination to participative politics. Moreover, there are not many other options at hand within democratic theory, if the purpose is to spread green values in an institutionalised manner. Stressing the absence of any guarantee of success is thus somewhat superfluous: it is the nature of politics to deny them. The politics of nature cannot escape the nature of politics.

However, likewise, green politics cannot escape the shortcomings of deliberative democracy either – shortcomings that concerns both the democratic *and* ecological aspirations of both radical democrats *and* environmentalists alike. The reasons why deliberative democracy is not the solution for achieving a sustainable society are, broadly speaking, twofold. On the one hand, the promise of the former cannot be fulfilled: deliberation does not work as finely as it is usually suggested, neither democratically nor ecologically. On the other hand, it is far from obvious that deliberation, useful as it is in some regards, should be institutionalised as a general pattern of social decision *without* fundamentally altering the outcomes provided by that very society in a wrong direction.

A first, unavoidable, limitation has already been suggested. There is no guarantee that ecological values will be embraced as a result of free and equal deliberations. However, deliberative democracy should increase the chances for green values to emerge, insofar as it makes more room for individual *change* as a result of the *exchange* of arguments with other individuals. This argument is based on the premise that the deliberative model goes beyond preference aggregation, to enhance the latter's discussion and transformation in the course of deliberation. Yet this is a problematic presumption. According to it, citizens *should* open their preferences to a process of comparison, discussion and, if rational valuation recommends it, transformation. Its normative character is undeniable: deliberative theorists *expect* such a transformation to take place. But if we pay attention to the current knowledge on practical reasoning and political motivation, we see that it is not reasonable to expect that citizens will dramatically change their preferences through deliberation (Johnson 2001: 222).

Furthermore, it is also the case that, when those preferences *do* change, it is not because a virtuous process of collective enlightenment has occurred, but rather because all kind of biases and unwelcomed effects have operated, so that the exchange *contaminates* instead of *rationalises* preferences. Informational pressure, social influences, amplification of cognitive errors, informational and reputational cascades, group polarisation, the over-weighting of common knowledge – all these group dynamics explain why deliberation does not result necessarily, not even usually, in wiser judgements and better outcomes (Sunstein and Hastie 2008). This affects the aggregation of information, to be sure, but also normative questions:

> If deliberation often fails to produce good answers to simple questions of fact, then it is also likely to fail to produce good answers to disputed issues of value. As it happens, the problems posed by informational pressure and social influences apply in all domains. They infect our most fundamental judgements about morality and policy, not merely judgements about facts (Sunstein 2006: 16).

This does not amount to denying that society is a collective enterprise, nor that therefore a "wisdom of the crowds" exists (Surowiecki 2004) that expresses the superior value of exchange between individuals. On the contrary. But it may be that an institutionalised deliberation is not the right setting for generating such wisdom, although it can perform a modest, supplementary function within a representative democracy and a liberal society. To put it differently, as we shall see later, group deliberation may not be the best way to put individuals in contact with each other, for them to exchange ideas and information. Besides, it may not be a *viable* way of doing so, since it is obvious that a mismatch exists between the overall number of people that constitute the *demos* and the people who can gather to deliberate once per week. Therefore:

> The great challenge facing such models of directly deliberative democracy – or participatory democracy, either– is this: how can we realistically involve the *entire* community of people affected by a decision in *meaningful* deliberations over that decision? (Goodin 2003: 2).

Everyone who has ever taught in a classroom is well aware of the problem posed by scale, time and the idea of a meaningful deliberation. Goodin himself is well aware of the impossibility of extending deliberation beyond a certain and modest limit, but at the same time wishes that its benefits could reach everyone. Hence his idea that democratic deliberation should be conceptualised "as something which occurs *internally*, within each individual's head, and not exclusively or even primarily in an interpersonal setting" (Goodin 2003: 7). In a similar vein, Andrew Dobson (2010) has underlined the importance of *listening* as opposed to just *speaking* in a democratic context, a virtue all the more important if we deal with voiceless subjects such as other living beings. These are indeed commendable goals. However, their fulfillment seem to depend on the degree of education, commitment and information possessed by individuals, rather than on any public measure aimed to *produce* reflective or attentive democrats. In fact, it is the lack of them, rather than the lack of the proper institutional settings, that weakens the case for deliberative democracy.

Of course, this is hardly a new point. The lack of information on the part of citizens has been often underlined as a major obstacle to radical forms of democracy. Whereas a representative democracy can cope with such shortage, a participative model could not (Sartori 2002). If citizens are to decide more or less directly, a minimum degree of *cognitive engagement* on their part is needed. Sometimes, in fact, more than that:

> To make sustainable choices, people must first understand underlying ecological foundations, i.e. that complex adaptive ecosystems provide the critical natural capital of essential goods and services and that the performance of these systems is the basis of our existence (Deutsch, Folke and Skanberg 2003: 353).

It may be argued that citizens are not necessarily oblivious to politics (Talisse 2004), which is true. But they might not be as interested as would be required by a deliberative democracy, nor seem so inclined to acquire the necessary amount of information for their judgements, to take Goodin's adjective, to be reflective. In the end, it is information that matters most: citizens who engage themselves in experimental deliberative arenas are firstly shaken by the new information they receive – it is only subsequently that the exchange of arguments with other people will benefit or, as we have seen, distort the collective decision. What we need most, in short, are enlightened people rather than deliberative groups.

Deliberativists are normally not inclined to give due weight to this crucial factor, perhaps because, again, they *expect* people to behave differently once they are given the institutional opportunity to do so. Yet it seems to me that people do not spend Saturdays in the mall instead of discussing public policy *because* there is no deliberative arena at hand – they are not, in a word, taking revenge on the representative system by seemingly wasting their time. They are leading the kind of lives they like, according to their personal preferences, which in turn derive from a mixture of education and circumstance. The more refined an individual is, the more prone will he be to behave as a reflective democrat that contributes – individually or in explicit agreement with others – to the general well-being and refinement of society. Thus the more inclined that individual also will be to reflect on value questions and to express a preference for some kind of sustainable society. Besides, crucially, this reflective and refined citizen does not contribute to a liberal, democratic and sustainable society just by deliberating or voting, but also primarily by *acting* in a number of realms, that is, by leading his life in a certain way. This is not just aggregation of preferences, it is something else. Therefore, deliberative forums can play a part, but not a central one, in such a society.

Now, we can see how unrealistic the green expectations on deliberation can be, if we take account of the fact that this general public is the same that is supposed to *integrate* and *listen to* nature. But even if this public were both able and willing to communicate with nature in order to bring her interests to a deliberative democracy, the task does not seem realistic. Natural signs are perceived and interpreted by mankind and discussed by people once they are incorporated into deliberative frames: otherwise nature's voice would barely be audible. It is humans who give meaning to the silence of nature. Human mediation does not work here as a simple chain of transmission: it produces a set of valuations later subjected to deliberation. That is to say, mediation is not just unavoidable, but also decisive inasmuch as it *creates* the claims of nature, since nature does not properly claim anything at all. After all, a social system cannot communicate with a natural system: communication is merely possible within social systems about those other natural systems (see Luhmann 1989). Democracy, as a discursive practice, is inexorably human. People can try to communicate with nature in their private lives, either individually or as members of associations, the result of whose actions can be incorporated into the political process in several ways. But the institutionalisation of human-nature communications does not appear to be advisable.

Still, such an emphasis in communication, rather than deliberation, points to a further problem in deliberative democracy. It is generally assumed that deliberative democracy is automatically inclusive. Yet it contains a strong potential for exclusion, because deliberation itself is an exclusionary practice. To deliberate means *at least* to build up arguments and to defend them through speech, persuading others of their value by rhetorical means. It is obvious that not all citizens have the same ability to do so (Hardin 1999: 116). It is simply impossible to provide all with equal "epistemological authority", so that everybody has the same opportunity to understand the issues at stake, develop a view of their own and convince others of the latter's preferability (Sanders 1997: 349). In the same way, as Iris Marion Young has suggested, deliberation rules are not neutral – assertive and confrontational discourses are much more valued than tentative, exploratory or conciliatory ones, revealing that the rule of the best argument reintroduces power into democratic debate, leading to an agonistic view of public sphere: "deliberation is competition" (Young 1996: 123). Hence deliberation adopts an elitist and exclusionary character for those individuals or groups with a lesser capacity for developing a given type of discourse.

This is an unsolved problem in deliberative theory – in fact, an unsolvable one. The imbalance of intellectual skills among people remains an obstacle to equal deliberation. In a representative frame, decisions are taken by a group of previously elected individuals, supposedly possessing similar epistemological authority. But how can the singular outcome of a debate in which some individuals dominate through rhetoric and persuasiveness over others be considered legitimate? Moreover, instrumental motivations and strategic attitudes are not necessarily excluded from deliberation, just because we believe that a pure communicative rationality is to prevail (Elster 1997: 17; Cohen 1998). Of course, in view of these difficulties, a more modest approach can be suggested, stressing *dialogue* rather than *deliberation* (see Escobar 2009). But this would in turn mean that a correspondingly more modest institutionalisation is granted to dialogic forums. On the other hand, democracy cannot merely be about communicating – it is part of a social system and thus it must lead to *efficient* decisions as well.

This is especially true of sustainability decisions. Like it or not, the relationship between expertise, democratic decisions and participation is not easy to arrange. It faces different problems with a same source: the deep divergence between technical and lay discourses. The dilemma is straightforward: whereas the legitimacy of any decision can be undermined in the absence of citizen judgement, leaving expertise aside might in turn lead to inefficiency. To put it differently, if citizens are to accept judgements made by scientists without being able to evaluate their foundations, is that compatible with individual autonomy and democratic polity? (O'Neill 1993: 5). To many, it is not. Hence the call for democratising science, that is, fostering some kind of participatory science conceived "as an instrument to dethrone science or to deprive scientific knowledge of its authority and legitimacy conferred by modern society" (Bäckstrand 2004: 109). A new civic science is supposed to spring from people's involvement that takes into account local and personal experiences. Deliberative instruments are key to its development.

Yet it is not clear what this science exactly amounts to, nor how useful it might be. It is not realistic to grant lay experiences or insights the same weight than experts: there must be limits to politicisation. This is not to fully dismiss public participation on scientific issues, but claims in favour of participatory science frequently express an exaggerated optimism in the will and ability of citizens to understand complex subjects in differentiated societies where expert knowledge is not easily accessible for the majority. This gap is the heart of the matter: citizens will have to accept expertise judgement in areas beyond their understanding. Specialisation is not a whim, but a key factor in human progress. Yes, sustainability is a process of social learning, whose implementation does require citizenship cooperation on a daily basis – hence better public understanding of science is surely necessary. Likewise, the institutional flexibility of deliberative procedures might help to achieve a greater balance between lay citizenry and expertise, between different understandings of risk and the need for rational and legitimate decisions on it. But this is on the condition that we do not carry it too far. Lay judgement on science does not lead to a new, more reliable science. That, I suspect, is wishful thinking. As it is the idea that everything can, or should, be directly democratised.

Deliberative Democracy and Open Sustainability

However, a firmer ground for connecting environmentalism and deliberation may be found, provided that the function granted to the latter is seen as supplementary and not central to social performing. This connection is supplied by an open conception of sustainability, according to which there is no such thing as a correct view of the sustainable society, but instead a plurality of approaches that should be free to be advanced and tested through both conscious decision and spontaneous arrangement. In this regard, the institutionalisation of deliberative arenas wherein different representatives (of citizens, experts, associations, governments) can discuss their views regarding particular issues (ranging from urban planning to infrastructure works and the setting of pollution standards) can contribute to the general social process of achieving sustainability. Such opportunities largely exist already, but remain mostly unnoticed. Their number should be enlarged, participation in them fostered. They can help to engage people in taking care of their surroundings, as well as developing a view of the good sustainable society, thereby creating a space where some versions of ecological citizenship can flourish.

It is on the local level that such arenas are most likely to be useful. That is partly the appeal of Bryan Norton's (2005) strategy towards sustainability: an adaptive management based on municipal deliberative processes. Norton suggests that municipalities ought to deliberate about the value of their environments, developing indicator measures for sustainability, and enforcing them. This would be a way of promoting resilience, according to Adger, insofar as the latter "requires flexibility and adaptation in decision-making on resource use and conservation" (Adger 2007: 86). Certainly, municipalities possess the key advantage of being close to the changing circumstances and conditions that a higher layer of governance may easily miss. However, that does not mean that an overall sustainability can be achieved this way, nor that a network of local deliberations is the most rational and effective means to do so. Rather it is, as deliberation itself, one of several means by which a liberal-democratic society develops an open view of sustainability.

But a problem remains. The belief that citizens in a deliberative context will spontaneously acquire ecological enlightenment and will push for green choices relies too much on an optimistic, naïve view of human nature, so frequently found in the utopian political movements. This is unlikely to occur. For all the praise of deliberative democracy, it would probably be easier for representative institutions to set sustainability as a social goal. As a general principle whose complexity affects all social spheres, accountable yet not directly democratic political bodies are more likely able to steer society towards that direction. Thus a combination of representation and deliberation would be more useful at the outset, fostering deliberation *within* representative institutions instead of relying solely on a would-be deliberative miracle.

As it happens, this might also be the incipient consensus among environmentalists. They now see deliberative procedures as supplementing, rather than replacing, reformed liberal democratic institutions (Carter 2007: 57). As Albert Weale puts it: "Representation, not participation, is the key problem of environmental policy making" (Weale 2009: 56). However, it is no coincidence that, as has already been noted, this retreat goes hand in hand with a rediscovery of the state – a green state – as an enforcer of sustainability. Of course, a capable state is indispensable for any politics of sustainability. But a green retreat back to state authority may also imply the wish to make sure that a strong sustainability is implemented. The rationale would be the following: if people's preferences are not so easily influenced through *talking* as they should be, then it is the state that must *act* through legislation and policy making.

A democratic state, though, cannot guarantee any given version of sustainability – unless a radical and homogeneous shift in the public opinion takes place. On the other hand, policy making is not, nor should be, either the only or the most important element of an open view of sustainability. On the contrary, environmental policies are the means by which the state sets those environmental standards that guarantee that society is sustainable in any given moment, without deciding the general picture that that very society is to adopt. In fact, once the basic principle is adopted, so that society becomes sustainability-orientated, policy-making should be considered as the outcome of a larger process of social learning rather than the beginning of it.

Democracy beyond Deliberation

The seduction that deliberation exerts on environmentalists has probably to do with the *explicit* way in which ideas and information can be exposed in the deliberative arena, because it seems more likely, for that very reason, that people will attend and concur to them. Hence the belief that radical social changes, as the one purported by classical environmentalism, have better chances to take place if we create the arenas wherein people can be convinced that they are wrong. This belief is mirrored in the contrary view that it is up to a green state to enforce sustainability – that is quite explicit too. But let us focus on democratisation, that is, on the question about whether sustainability can be democratised. The answer is that it can, but not in the conscious, institutionalised manner that most environmentalists aim for.

An open conception of sustainability is democratic in the sense that it is carried out via the participation of several actors in a number of ways. This includes the participation through the market, the informal deliberation in the public sphere, and the communication struggle that social movements engage in for a definition of the good life. An enormous and complex system of debate thus emerges, in which governments, citizens and collective actors simultaneously *act* and *communicate* around sustainability. This entails a wider conception of democratic participation, since it includes both conscious and spontaneous, direct and indirect forms of contributing to sustainability. It is not just centred on voting or deliberating, rather it includes informal conversation and all kinds of exchanges as potential ways of influencing sustainability.

But why is sustainability not to be directly democratised? The reasons abound, but they might be less clear for those who believe that the ecological crisis is just the visible tip of a wider human malady – a crisis of civilisation that threatens survival and demands a thorough new start for society. But it is not so. Humanity has bettered her condition steadily throughout history and continues to do so through fiercely human means: innovation, exchange, reciprocity. The current need to reorganise socionatural relations is but one more proof of human's ability to correct itself – a proof of adaptive capability. In this regard, environmental risks are the other side of material progress. But we could also say material progress is the other side of risk. A material progress that is valuable, despite the fact that moral progress may advance more slowly. Although it is not so slow, either: torture can be still practiced, but it is almost universally rejected on ethical grounds. Therefore, we should not embrace a sinister portrait of modernisation just because it happens to produce new threats. Of course it produces them; as well as benefits. To someone who adheres to a radical rejection of modernity, this argument is radically flawed. But let us take for granted that modernisation is to be more carefully watched. Yes, just *watched*, because it is not clear at all that it can be *controlled* without actually being *stopped*.

It is well known that environmental risks are not so easily reducible to particular decisions. Some of them are, but many others are not. Modernisation and development are the outcome of countless decisions that produce together an autonomous process of innovation and change. Yet that is not equal to a lack of control: there are regulations, courts, environmental standards. But the point is that we cannot simply submit this enormous, self-replicating, complex entity to political control through the deployment of a few citizen assemblies – nor by the planning of a centralised state. It does not work either way. The whole foundation of our wealth would crumble. If we wish to unleash a revolution, that is fine, but if we want to keep the advantages of advanced societies while making them sustainable, a dose of realism is advisable.

What does that amount to? It means that sustainability cannot be *fully* democratised. We cannot politicise every source of environmental risk-creation without leading society to a dead-end. Thus it is better to concentrate on those spheres and moments of risk-creation and risk-management where the application of deliberative principles and structures make sense – as it is the case with the local impact of public or private projects or with the governmental regulation of environmental standards. However, such a limitation on *direct* politicisation leaves plenty of room for the *indirect* democratisation of sustainability. This is an important point. Debating and deciding within institutional structures is not the only way in which citizens can give shape to reality. Their decisions in the market and in the household, as well as their personal behaviour and their informal conversations, equally contribute to that shaping. Indirect democratisation must be understood as *influence* rather than as *decision*. In other words, our decisions influence the daily negotiation of reality, although they do not directly decide how reality turns out to be. And the same goes for collective actors, such as social movements or even corporations.

How can this work? The key point, I daresay, is the relationship between information, decision and outcomes. How can we connect people's decisions in a way that enable them to acquire information through the exchange of ideas and goods with others, thus making the final outcome a fine reflection of a collective intelligence at work? Both deliberativists and environmentalists stress the need to *personally* connect people and make them *speak* with each other for a transformative communication to take place. It is in this way that reality is supposed to change for the better:

> These changes can neither be prescribed nor logically deduced and implemented by decree. They rely on communicative and participative procedures that help all participants to define their known vision, discuss common strategies, and find closure about what needs to be done (Renn 2009: 252).

There is an obvious preference for institutionalised procedures and concerted communication over informal systems of interaction and exchange. This also entails a preference for a *guided epistemology* in which actors deliberate explicitly about the preferred outcome, in opposition to a non-institutionalised process of decision through spontaneous interaction between people that both *communicate* with each other and *act* individually. This *unguided epistemology* is, of course, the sort championed by Friedrich Hayek, whose subtle analysis of the conditions for social evolution underlined the importance of spontaneous cooperation and non-verbal actions and events:

> The intellectual process is in effect only a process of elaboration, selection, and elimination of ideas already formed. And the flow of new ideas, to a great extent, springs from the sphere in which action, often non-rational action, and material events impinge upon each other (Hayek 2008: 32).

It is human interaction and exchange that makes material progress and cultural change possible and thus sustainability too. Although it is frequently assumed that the picture of the individual that derives from this description of reality equates that of a selfish preference-maximiser, this is by no means true. Rationality is considered by Hayek in social terms, as the by-product of a social learning process wherein key institutions such as language, money, and the market, but also the law, are spontaneous orders that result from human action but not from human design. Social order would then be the result of adaptive evolution. It is a super-individual process that operates through an ever-increasing individual specialisation, so that every person can take advantage of the others' knowledge without having to acquire it herself: a collective brain instead of an individual one (Ridley 2010). It is thus an unguided rationality.

Normally, we identify this kind of epistemology with the market. Not without reason, since the market is a site for human interaction and exchanges which largely determines the shape of society. Markets are not perfect, to be sure, especially when it comes to common goods, but they are very efficient in allocating resources, fostering innovation and spreading the latter's fruits. They are the product of human interaction, without any magic working at all. As Von Mises put it:

> There is nothing inhuman or mystical with regard to the market. The market process is entirely a resultant of human actions. Every market phenomenon can be traced back to definite choices of the members of the market society (Von Mises 2007: 258).

Yet these members, trying to maximise their preferences, do not always behave rationally, as neo-classical economy suggests. Insofar as individual choices are not made in a social vacuum, peer-pressure and status anxiety can play an important part in decision-making. Besides, not everybody makes the necessary effort to collect the available information or pauses enough before satisfying a preference. However, that does not suffice to deny the general effectiveness of markets as vast systems of information exchange that also happens to produce knowledge in a way that maximises the ability of individuals to do so. It is the *interaction* that makes the difference – and the greater the interaction, the greater the difference.

Mark Pennington has thus distinguished between two simultaneous functions performed by markets: on the one hand, the market is a spontaneous coordinating mechanism that enables people to adjust their behaviours through the price system, whereas, on the other, it acts as a social medium for the discovery *and* communication of knowledge, that is, "an inter-subjective *learning* procedure in which contradictory ideas widely dispersed amongst individuals and firms are constantly tested against one another, and where new values are discovered and disseminated via a process of emulation" (Pennington 2003: 727). Thus we can talk of an *"epistemologically productive* market interaction" (Foster and Gough 2005: 100).

Markets, therefore, have to do with knowledge and information as much as with goods and services. For instance, the preferences shown by a given community (themselves a reflection of the level of education and refinement its members possess), will shape both its commercial landscape and its labour supply in a certain way, and not in another. All kind of exogenous factors, of course, can distort this outcome. Public or private monopolies, rigid labour laws, absence of reliable courts, and so forth. But given some basic conditions, the premise applies: a society acquires a given shape depending on the outcome of the interactions between its members.

Nevertheless, there is no reason to restrict this explanation to markets. When we refer to markets, we allude to an exchange of goods or services, whichever its kind, orientated by the price structure, whose side-effect is the production of knowledge and the emergence of values. But outside the markets citizens and collective actors engage themselves in a network of communications – through words as well as deeds – not orientated by prices. These communicative actions can be conscious or unconscious: they can have a goal, as exerting cultural or political influence, but they can also be just a part of the everyday life, like behaving in a certain manner. Exchange, emulation, reciprocity, cooperation, competition: everything is at play in this web of actions and meanings. At the end, though, the common denominator is human interaction. This is epistemologically productive as well, whatever the quality of the final outcome. Yet all actors exert a given amount of influence on the latter. In this regard, everything is political, insofar as every action impinges on society's shape.

Yet it is incongruous to disentangle markets from non-priced human exchanges. It is not, as conventional wisdom has it, as though there is a separate realm for economics where people behave as maximisers of preferences in opposition to a non-economic realm where *those same people* behave in a virtuous fashion. These spheres overlap, as our behaviours overlap. Pennington again: "Markets already *are* a part of the 'public sphere', where preferences and values are constantly shaped and reshaped as the process of competitive experimentation unwinds" (Pennington 2003: 737). Thus the important thing is the nature and quality of our interactions and exchanges. Therefore, the most relevant condition for the social process is the way in which the different members of society are related to each other, i.e. the easiness and quickness of the communication channels that are available to them.

Sustainability beyond Democracy

Let us now return for a moment to Robert Goodin's remark about the difficulties faced by any attempt to put deliberative democracy at work. He asks: "how can we realistically involve the *entire* community of people affected by a decision in *meaningful* deliberations over that decision?" (Goodin 2003: 2). This, as Goodin himself acknowledges, is unfeasible. It is a matter of scale, to begin with. But we have already ascertained that just gathering people in a deliberative arena for them to jointly decide about some given issue does not necessarily lead to a good decision, let alone to a decision that maximises the knowledge and abilities of the group's members. Therefore, it is not just scale, but also human psychology. Surowiecki has remarked that "the more influence a group's members exert on each other, and the more *personal* contact they have with each other, the less likely it is that the group's decisions will be wise ones" (Surowiecki 2004: 42; my emphasis). Hence a key to better decisions lies as much in non-institutionalised processes such as the market or informal conversations as in non-personal networks of decision *and* concerted action.

We are not making explicit decisions all the time; rather a lot of decisions are made on the move, in the very process of living. This looseness, however, appears to hinder the kind of preference transformation pursued by environmentalism: a way of influencing people's wants to make them less individualistic and materialistic (Carter 2007: 65). But that is not necessarily so. An explicit pressure on people to behave differently is, again, not very effective. Apart from those basic mandates that are translated into law, sustainability cannot be minutely decreed – let alone an openly conceived one. This is why Goodin appeals to inner reflection, whereas Foster adds to it a super-individual process of spontaneous social learning:

> This doesn't require that judgement be *unguided*, but it does mean that guidance can only come from criteria internal to the idea of judgement itself. (...) The collaborative attention and overall, on-balance disinterestedness which has to inform the collective judgement taking us heuristically forward is what we might call *social intelligence* (Foster 2008: 142).

The more interconnected the individual intelligences are, the quicker will this social intelligence advance towards a sustainable society. It belongs to the very nature of social learning, though, that the goal to which efforts are headed remains relatively open – as is the case with a *general* conception of sustainability. A society that pursues an open sustainability is by definition a free one wherein different approaches to the latter are tried in different realms, from individual behaviour to business, from legislation to science and technology. A renewed environmentalism thus acknowledges that "creating, trying out, rejecting, or modifying and improving, and trying again to find new ways of moving this species toward a socially just and ecologically sustainable way to occupy this planet is the only game big enough for all of us to play" (Stoner and Wankel 2008: 3). There are signs of just that. Whereas Jonathon Porritt (2005) has suggested reviewing the business case for sustainability, as opposed to the traditional green suspicions about capitalism's adaptability, Stewart Brand has proposed replacing the precautionary principle with a more innovation-friendly alternative:

> The emphasis of the vigilance principle is on liberty, *the freedom to try things.* The correction for emergent problems is in ceaseless, fine-grained monitoring, which largely can be automated these days via the Internet, by collecting data from distributed high-tech sensors and vigilant cellphone-armed volunteers. (...) Managing the precautionary process in this mode consists of identifying things to watch for as a new technology unfolds (Brand 2009: 164).

Be that as it may, this circulation of ideas and projects runs parallel to a public conversation about the kind of society – and hence the kind of sustainability – we wish to have. More exactly, though, action and conversation overlap and intertwine; in the end, they are the same thing. Thus it is not simply sustainability *through* business, as ecological modernisation theories have it, that is suggested. We talk of a democratic *society* rather than of an all-encompassing democratic system, whereas, at the same time, this democratic component goes beyond a simply business-orientated society.

If, therefore, we are considering how people are connected, in a way that enables them to communicate their preferences and ideas, as well as exchanging or sharing them, setting up collaborative projects, and so on, it would be absurdly romantic to dismiss the increasing role that the new information technologies are playing in making that possible in an unprecedented way. Therefore, it is out of old-school reverence for citizen assemblies that communication technologies are not properly appreciated for what they are: a formidable device for facilitating exchange and collaboration between people that otherwise would have remained separated or disconnected. This is happening on every level, from corporations to governments, at a different pace, but it is obvious that a deep societal change is taking place:

> These changes, among others, are ushering us toward a world where knowledge, power, and productive capability will be more dispersed than at any time in our history – a world where value creation will be fast, fluid, and persistently disruptive (Tapscott and Williams 2008: 12).

This is democratisation beyond democracy, but maybe not beyond deliberation, if we take deliberation to be more informal and loose than the one required by the ideal speech situation. People who collaborate via the internet or with the web's aid are not just deliberating, but also acting – yet they deliberate as well, and by acting in a certain way they are also engaged in a communication act: they are saying something about the things they like, offer, or wish to change. It is truly a separation of powers if there ever were one! Some go on and predict a whole new world:

> In other words, we may soon be living in a post-capitalist, post-corporate world, where individuals are free to come together in temporary aggregations to share, collaborate and innovate, where websites enable people to find employers, employees, customers and clients anywhere in the world (Ridley 2010: 273).

Why not? Marx himself predicted the melting of a *successful* capitalism into some kind of wealthy socialism. It could even happen that this new *ethos* make some reforms possible, e.g. designing markets for natural capital or creating new currencies, that gives shape to some kind of "polycapitalism" in tune with a sustainable society (Spen Stokes 2009). But it is not necessary to go that far to welcome the new communication tools and the benefits they are bringing about. The epistemological qualities of exchange and interconnection are already being sharpened by the increase in the number and easiness of interactions between people across the world: a collaborative society is emerging.

This all includes, needless to say, exchange and collaboration *towards* sustainability. Society is already shifting in the direction of a public – albeit loose – commitment to a sustainable reshaping of socionatural relations. It is thus to be expected that an increasingly open society meets open sustainability.

To be fair, this is already happening. On the one hand, new technologies are making easier and easier to collect more and better information about the natural world and the human interaction with it. It is obvious that we owe our knowledge on climate change to data gathering and computer modelling (see Edwards 2010). On the other hand, though, Edward O. Wilson (2003) has praised the ongoing creation of an "encyclopaedia of life" that put all the information that we get on species already known into a single great database, with a page that's indefinitely extensible for each species in turn, and that would be available to anybody anywhere, on demand, and free. It is easy to see the advantages that *sharing* data entails in the natural sciences, where the sheer scale of the problems themselves could not been matched *until now* given the scale of people trying to solve them. Likewise, Stewart Brand refers to Dan Janzen's project of fostering a "Barcode of Life", that is, a handheld device able to give anyone the genetic information of any living being that has already been classified, making ecology a predictive science by identifying *all* species. As Brand notes: "Just as pocket calculators democratised math, DNA bar-coding makes the whole world bio-literate" (Brand 2009: 271). Without science and technology, Arcadia cannot exist.

Outside the sciences, however, information technologies and the kind of collaboration and exchange they facilitate are also contributing to the transition to sustainability. Sometimes, new practices are not explicitly orientated towards sustainability, but contribute to it by saving resources or – seeing it differently –

maximising the efficiency of human exchanges. An example at hand is that of "collaborative consumption" (Botsman and Rogers 2011), instances of which are sites such as Freecycle, eBay, Couchsurfing and Zipcar, that allow people to share or re-sell instead of just buying a brand new product. More specifically, as Tomlinson (2010) summarises, information technologies provide information that encourage people to go sustainable, improve the efficiency with which we enact our daily lives, provide opportunities to address environmental issues through improved collaboration and, on the institutional level, help to improve infrastructures (smart grids, for instance), large-scale planning and policy decisions. Simultaneously, the primacy of information is restructuring processes, institutions and practices of environmental governance, giving way to a so-called "informational governance" (Mol 2008). Needless to say, conventional forms of policy and governance are not replaced overnight by the new ones, rather they cohabitate and mix. And the same goes for non-institutional practices and social ways of interacting.

Social movements are a good instance of that mixture. In the last decades, they have operated more and more in the communicative realm of society. It is not only that they have taken advantage of the digital tools as organisational resources that alleviate their costs and connect their members more easily. Above all, they have focused on communicative actions that try to influence public opinion and governments. They try to change the dominant social frames, spreading alternative codes and practices, hence fostering cultural change (see Melucci 1996; Snow 2004: 384). Their politics of persuasion take place in the public sphere, where it reaches citizens, politicians and corporations. In our society, this war of meanings takes place in the communication network. There would be, in fact, a potential synergy between the rise of "mass self-communication" and the autonomous capacity of civil societies to shape the process of social change (Castells 2009: 303).

Thus direct democratisation is not the only tool available to citizens *if* they wish to exert some degree of influence on sustainability. Naturally, they could as well not be interested. But although a majority of them may not care, it is enough that a bunch of more active innovators – together with governments – create the conditions for the rest to join them in *practicing* sustainability, making ecological virtue just another feature of everyday life. Remarkably, the models that are emerging around the new tools provided by the information technologies rely less on central control and more on the self-organising of a critical mass of people and organisations that initiates small experiments and social innovations that can lead to pervasive changes in societal behaviour (Tapscott and Williams 2010: 18). Of course, citizens do vote and hold political opinions, but, above all, they lead lives. Some would say that they are *passive* bearers of the lifestyles proposed by something called "the system". But that amounts to ignore the essentially *dynamic* quality of society, as well as the *active* role that citizens play when they make the countless decisions a normal life requires.

Ecological consumption is an instance at hand. It entails the application of environmental criteria when choosing goods and services in the market (see Seyfang 2006; Jackson 2007). That includes avoiding polluting brands and choosing those with a clean environmental record instead, but also, more interestingly, making a whole range of wider choices that lead to a sustainable lifestyle. By sustainable, it is noteworthy, we must not exclusively understand frugal or natural-friendly, but *just* sustainable. An open view of sustainability does not pre-determine the basket of goods and services that qualify for a consumer to be "ecological". It is up to her to decide, provided that she makes *some* contribution to the kind of sustainability she chooses to support – in other words, provided that she does not consume in an unsustainable way. After all, markets are information systems; our consumer choices send signals about the preferences we have, that is, about the society we wish to live in.

Although these may all seem small changes, especially for those who believe that only a radical restructure of society is acceptable as a path to sustainability, their aggregation in the long run is not. Besides, the transition to sustainability is incremental rather than rupturist. For all the talk of leaps and tipping points, that is not how sustainability will be achieved. It is a transition, not a rupture. Furthermore, crucially, it is not a transition to some kind of steady-state society, a point in time where humanity stops to change itself and its environment. That is never going to happen. In this regard, we have to *free* sustainability from environmentalism. Or, to put it the other way around, we have to renew environmentalism, for it to embrace an open, technology- and wealth-friendly approach to sustainability. An approach that is reached via trial-and-error rather than by decree.

In sum, democratising sustainability means something else other than just institutionalising arenas for citizen deliberation – although such measurements do not need to be excluded. It includes market practices, civic action, collective mobilisation, business innovation, and governmental supervision. It involves communication and action, both rational and extra-rational. An open sustainability needs, in sum, as many insights and sources of change as possible.

Chapter 8
Ecological Citizenship and Sustainability

The Green Turn to Citizenship

Although several liberal institutions have been subjected to the green critique, this chapter will solely focus on the notion of ecological citizenship. Not only because it is the most recent and promising development within green political thought, but also in view of the undeniable advantages it possesses when discussing liberal sustainable society. In fact, the reflection on ecological citizenship is central to any meaningful discussion about a democratic sustainability that goes beyond deliberation and into the very fabric of society in order to foster a collaborative social order. After all, citizenship is an institution located between the formal political organisation and the informal realm of civic action, as well as between the public and the private. It is thus able to reflect the nature of a liberal social order while serving simultaneously as a tool for its transformation. This is especially relevant in connection to environmental issues, since they are known to be present on both an individual and collective level, in a way that makes achieving sustainability dependent on public as well as private actions. It is thus also related to the market, as the arena where citizens make decisions said to be private but ultimately decisive for the state of the environment. In sum, it has to do with people's motivation to behave – or not – in a sustainable way and the instruments they possess to do so. And it is noticeable that empirical studies are starting to appear that try to test the hypothesis of the ecological citizenship in the real world, hence providing valuable insights about the real conditions of a realistic sustainability. Therefore citizenship helps us to think about the relationship between sustainability, liberal democracy, and the moralisation of the socionatural relationship.

It is even possible to identify a "turn to citizenship" in green political thought, such is the enthusiasm with which the idea has been welcomed (see Dobson 2003; Dobson and Valencia 2005; Dobson and Bell 2006; Smith and Pangsapa 2008). Of course, this turn is, first and foremost, a shift of attention in the direction of the individual as a decisive factor – if not the decisive factor – for achieving sustainability and spreading a more ecocentric worldview. But it is also a testimony to the adaptability of citizenship itself. Whether we embrace or not the famous analysis by T.H. Marshall (1987), according to which the emergence of new types of citizenship – defined as the body of rights and duties that accompanies the full belonging to a given society – has run parallel to the attribution of new rights to individuals, it is obvious that the institution of citizenship is intrinsically dynamic and part of an open-ended political process. Such flexibility gives credit to the suggestion that an ecological dimension can be added to the civic, political and social ones previously incorporated into the content of citizenship. From this standpoint, ecological citizenship would simultaneously be a correction and a supplement to the classical understanding of citizenship (see Steenbergen 1994: 142; Smith 1998: 98).

But why citizenship *for* sustainability? Well, besides the obvious fact that a theoretical machine like the green one cannot stop and ends up devouring everything, a single fact should be underlined, namely: a citizen is a person. A trivial statement, to be sure, yet it embodies all the possibilities, as well as the complications, intrinsic to the very idea of an ecological citizenship. It is less trivial to say that we are currently almost 7 billion people on the planet, and rising. To make this global society sustainable, changes are required on a structural level, but also on an individual one. After all, people lead lives that oscillate between the public and the private realms, lives that are led *everywhere*, so that the domestic realm acquires a global reach with enormous implications for sustainability:

> This shift is taking place partly because today's most challenging environmental problems develop independently of geographically and politically defined borders. Thus, without a global human collective characterised by an outspoken willingness to make behavioural and attitudinal changes, no individual state – whether democratic or not – can succeed in solving environmental problems such as climate change (Jaggers 2009: 19).

In short, the truism goes like this: there is no sustainable society without sustainable citizens. But, at the same time, the attitudes of these citizens, not just their behaviours, should shift on a deep level for a strong enough sustainability to be achieved (Dobson and Valencia 2005: 1). The whole project of an ecological citizenship thus entails an educative process at the end of which individual people behave differently *because* they develop a different attitude towards the environment. Whereas the market and democracy have to do with individual preferences, ecological citizenship, it seems, points rather to people's *habits* in their everyday life – that is, to those individual actions that are more or less automated and do not depend on particular decision-making processes, be they individual or collective. Of course, habits are the expression of behaviour, and behaviours are the outward expression of inner attitudes.

In turn, though, attitudes depend on the consciously adopted or – this is important – somewhat unconsciously followed conception of the good or personal life-plan. It is obvious that the proliferation of an ecological citizenship would almost instantaneously lead to sustainability, or, furthermore, to a deep green sustainable society. Hence the particular appeal of citizenship for the green programme. However, it is not so obvious how such proliferation can take place *without* developing a massive moralisation project that actually changes people's deep attitudes and does so interfering in their life-plans. Remarkably, environmentalism has turned to citizenship while simultaneously abandoning its classical anti-statist stance, assuming instead that an ecological state is key for delivering an ecocentric society (see Eckersley 2004; Barry and Eckersley 2005). But precisely because people are real, they are not so easily changeable and can often be stubborn. This would explain the obvious contrast that exists between the *conceptual* and the *sociological* ecological citizen – between the desired and the real one. We will come back to this. The point is that maybe environmentalism is misunderstanding the nature of citizenship, taking it as the starting point of a moralisation process, instead of contemplating it as the legal culmination of that process. But then again, it could be retorted that ecological citizenship differs from the traditional interpretation of citizenship, and should not be weighed in the same way.

Defining Ecological Citizenship

It has been rightly pointed out that there are competing views on what ecological citizenship consists of, rather than there being an univocal model for it, meaning that ecological citizenship would actually remain "under construction" (see Valencia 2005). However, there is a dominant theoretical blueprint that serves as a basis for discussion – namely, the one brilliantly developed by Andrew Dobson (2003). Although it is not the only conceivable form that ecological citizenship can adopt, there are not so many possible variations on it *if* this concept is to contribute to the moralisation of the socionatural relationship and to the achievement of a strong version of sustainability.

The most salient feature of ecological citizenship is the shift from an emphasis on rights to an emphasis on duties and responsibilities. Rather than as a holder of rights, the citizen is to be considered as a bearer of duties towards the environment. This entails a shift for the politics of sustainability, since a focus on attitudes and lifestyles involves the – mostly correct – belief that cultural change in the direction of sustainability does not germinate in the institutions, but on the civil society and the private realm, influencing governments in turn. But this is not to suggest that people change in isolation. It is to recognise that individual actions – whether they take place in the public space, the supermarket, or at home – can be as influential on an aggregate level as the political participation in the institutional arenas, if not more. As I said, I am not sure that citizenship is the right container for such content, but I will elaborate on this later.

Such moralisation of the public life goes hand in hand with the politicisation of the private realm. This is a second distinctive trait of ecological citizenship: the relevance it grants to the private lives of citizens. If feminism proclaimed that the personal is political, environmentalism adds that the domestic is ecological – and henceforth political. In the words of John Barry:

because the private/domestic sphere is ecological (in the sense of being a site
of ecologically significant behavior and action), it is by the same token, from a
green/ecological point of view, political (in being a site for "green citizenship"
and legitimate state regulation (Barry 2002: 148).

Thus the domestic behaviour of citizens is not their own business anymore: it
is everybody's business. It is in the public interest that such behaviour becomes
environmentally friendly. Furthermore, it is a global business. Ecological
citizenship is not nationally or regionally constrained, for the same reason that
ecological damage is not either. The arena in which the ecological citizen is to act
is *produced* by their material relationship with the environment. Think globally,
act locally: the green approach to citizenship expresses a global orientation that
questions the traditional, nationally-bounded, understanding of citizenship (Carter
2001: 5). It is plain that our local actions can affect the global environment,
especially on the aggregate level. One citizen's overconsumption of energy is a
statistic; a billion is a tragedy.

This new relationship between the local and the global is mirrored by the
new relationship between the single individual and the global environment,
because it turns us into "citizens of the Earth" (Dobson 2000b) or "citizens of the
environment" (Bell 2005). Traditionally, citizenship has represented a relationship
between the individual and the state, in which the two are bound together by
reciprocal rights and obligations. Thus ecological citizenship proposes a new kind
of bind, wherefrom a new kind of duty emerges, namely, the duty to reduce the
occupation of ecological space, and doing so as a matter of justice (Dobson 2009:
135). For all its potential faults, the ecological footprint is meant to express this
obligation: "The relevance of the ecological footprint to us is that it contains the
key spatial and obligation-generating relationships that give rise to the exercise of
specifically citizenly virtues in the environmental context" (Dobson 2009: 135).
We all produce a footprint; we all can do whatever we choose to do in order to
reduce it; we can reduce it significantly leading a more sustainable lifestyle. The
ecological citizen could then be described as someone who is morally committed,
out of justice, to reduce their ecological footprint.

The Challenge to Liberal Citizenship

It is frequently assumed that insofar as ecological citizenship is grounded on the
recognition that we have moral duties towards the environment that stem from
our material relationship to the latter, the liberal model of citizenship is to be
transcended for a sustainable society to emerge. From a green point of view,
liberal citizenship would simply be orientated to the satisfaction of preferences:
whereas citizens demand their rights in the public sphere, they fulfil duties and
participate actively in their private one, in order to achieve their life-plans. That
would amount to a *strategic* view of citizen belonging. In such a context, duties

and obligations derive from a contractual logic, so that the fulfilment of any other duty can just be attributed to an expression of *care* (Fraser and Gordon 1994: 101). The green approach to citizenship draws openly on the feminist critique. As a result, the private orientation and the contractual foundation of liberal citizenship are openly challenged.

As a matter of fact, the novelty introduced by ecological citizenship does not reside so much in the proclamation that citizens do have duties, but rather in the source and beneficiaries of these. The whole idea of ecological citizenship is rooted in the relationship between the individual and the natural environment. There is no possible reciprocation in such relationship, no contract to be, albeit metaphorically, signed. And the same goes for the future generation of humans that may also be considered a source of obligation. We can expect *nothing* in return from our environmentally virtuous behaviour – apart from a moral satisfaction. Therefore:

> The liberal democratic contractual relationship between individual and state – while still in place – is in the ecological model of citizenship complemented and even overshadowed by a moral relationship between individuals regardless of where they live on Earth (Martinsson and Lundqvist 2010: 519).

It would then seem that ecological citizenship must be grounded on a different social ontology than its liberal counterpart: non-reciprocal duties, non-contractual obligations. Andrew Dobson has elaborated on this:

> An alternative social ontology of material embeddedness calls into question the "common sense" reciprocal expectations of morality to be found in the contractual idiom, and opens up the possibility of alternative models of citizenship relations (Dobson 2003: 50).

Thus the cultural mythology underlying liberalism would preclude the recognition of a non-contractual reciprocation, let alone one that has to do with non-human subjects. Admittedly, there is a model of citizenship that takes citizen duties to its very core: the republican model. Individuals are therein integrated in the community as a whole, the social and personal identities being actually formed only on the basis of shared traditions and inter-subjectively recognised institutions (see Habermas 1996: 22). Although it is doubtful that the social conditions for reaching such – probably romanticised – republicanism exist anymore, there is a further reason for its rejection on the part of environmentalism, namely, its contractual basis. As Dobson claims, greens do deny that citizenship must be contractual *by definition*, defending instead unilateral and non-reciprocal citizen duties, determined by their material embeddedness in the natural environment rather than by their personal autonomy (Dobson 2003: 46–50).

It is remarkable that this close association between citizenship and duties provides ecological citizenship with a strong pedagogic dimension. In the context of a society orientated to sustainability, ecological citizenship would become a practice that enables the gradual learning of those actions and omissions needed to achieve the former. Henceforth, citizenship is considered a type of social learning that contributes decisively to the socialisation of green citizens and the creation of communitarian bonds. In fact, this usage of citizenship is related to an old aspiration of classical environmentalism: the creation of a sustainable community along Arcadian lines. Long before the idea of an ecological citizenship could appear, O'Riordan (1976: 315) had suggested that citizenship is a moral, social, and political attribute without which a community grounded on a strong understanding of the public and the commons is not feasible. In this vein, it is claimed that the citizen protection of natural systems and landscapes can only take place through a direct participative relationship between the local community and the environment it inhabits (Light 2001: 28). Thus the *personal* relationship between the citizen and the environment leads to a *social* relationship between the community and the environment, and therefore the ecological citizen becomes the means for achieving an ecological community.

However, there is a further dimension of ecological citizenship that must be taken into account. It is no other than its *political* dimension, which is partly related to the traditional green request that liberal democracy be replaced by a more participative model, or at the very least that representation be strongly supplemented by more direct – normally deliberative – decision-making processes. A deliberative green democracy cannot survive without engaged ecological citizens, eager to discuss and decide on sustainability itself. But I say partly, because ecological citizenship is also related to a version of contemporary citizenship that sees civil society as a global arena wherein the *real* needs of people, as opposed to the *spurious* interests of the beneficiaries of the capitalist system, can be vindicated. Thus the global civil society becomes the arena where a democratisation takes place *against* and *apart from* the state (see Dryzek 1996; Christoff 1996). In other words, the globalisation process and its impact in civil society would seem to open up new possibilities for the transition to a sustainable society (Valencia 2005). And vice versa: ecological citizenship should be a tool for politicising nature as part of a broader political struggle that aims for a democratisation of citizenship itself (Latta 2007). This sort of citizen insurgence, whose momentum is admittedly receding as of late, is definitely anti-liberal, so that the kind of sustainability it promotes is closer to a strong than to an open version of the latter.

Of course, it is certainly not compulsory that, were ecological citizenship to be implemented, liberal citizenship had to be displaced for that reason. They could coexist, since the body of rights and liberties associated to the latter would hardly vanish. But it is also plain that an *institutionalised* ecological citizenship would be more than a mere supplement to the liberal one – if only due to the amount of restrictions that it would impose to personal autonomy and the corresponding realisation of life-plans, in the name of a deep green view of nature and a strong version of sustainability.

Sustainability, Morality, Citizenship

Is thus ecological citizenship a consistent foundation for building up the sustainable society and renewing green politics? The answer, it seems to me, is negative. Or to put it differently, it is negative unless we change the way in which we approach this ambiguous institution. Because, for all the appeal that non-reciprocal moral duties may possess, it is not the relationship between the citizen and the natural environment that should be stressed – but the one between citizens and sustainability. Nature, after all, is gone. If we insist on moralisation as the main motive behind the greening of citizenship, its reach can be paradoxically diminished.

We can begin by asking what exactly the content of ecological citizenship is. Now, this is an interesting point to make, because, insofar as none of the duties towards the environment that a citizen *should* bear is made legally binding, such content is nil. Of course, it could be argued that this is precisely the idea: ecological citizenship stresses moral duties freely assumed by self-conscious citizens. In other words, then, its content is *practical* – it is made of people's actions on behalf of the environment on a basis of justice. If, as Bryan Turner (1994: 159) puts it, citizenship can be defined from a sociological point of view as the set of practices that constitute individuals as members of a community, ecological citizenship consists of and is produced by such practices, insofar as they reflect a sense of moral duty towards the natural world.

Yet this citizen is hard to find. As a consequence, there is an obvious contrast between the normative proposal for an ecological citizenship and its sociological reality. Again, it might be replied that such is, precisely, the idea: to create a blueprint for the *future* ecological citizenship that stresses the moral duties of the *current* citizens towards the environment. But it does not seem wise to ignore reality, or, we could even say, to *overburden* it with an unfeasible theoretical proposition. The fact remains that such a citizen is nowhere to be seen, nor will be easily generalised. Although the empirical studies that search for this particular kind of citizen are still scarce, their conclusions are not encouraging, whether they refer to the predictably careless Spain or to the environmentally committed Sweden (Valencia, Arias, and Vázquez 2010; Jaggers 2009; Martinsson and Lundqivst 2010). In each and every case, the problem remains the same: a formidable gap between declared *intentions* and *behaviours*, that is, between what is *said* and what is *done*. This comes to demonstrate that there is no one-way relationship between values and action, but also points out the social, rather than individual, character of both values and practices (see Hards 2011). That is why the "willingness to act" that Jaggers proposes as criteria for identifying an ecological citizen may be insufficient: talk, after all, is cheap. Still, such willingness may suffice if it is accompanied by a reasonable degree of action, no matter what exactly the *reasons* are that motivate it.

This means that, if we lighten the requirements for what counts as an ecological citizenship, the outlook may not be so gloomy. On the one hand, because although it is dubious that a *deep* green virtue can be generalised, more *ordinary* green virtues, that lead to a similar outcome for the environment, may well spread (Martinsson and Lundqivst 2010: 532). But, in addition to that, there are normative reasons for such alleviation. They have to do with the need to define sustainability in an open way, as well as with linking ecological citizenship to the practical articulation of the former in the social arena.

It has already been stressed that the non-reciprocal duties towards the environment that constitutes the core of the strong interpretation of ecological citizenship are essentially moral duties. They are neither reciprocal, nor mandatory; besides, their lack of fulfilment is not punished. They belong to the realm of virtue. A virtue can be loosely defined as a *behaviour* orientated to the good that is *not* legally enforceable. If it is legally enforceable, then it is not virtuous. Moral duties are to flourish "in the interstice of rights", because the state is not the only source of our moral obligations (Nozick 1984: 503). Or it shouldn't be.

In this regard, ecological citizenship can be said to belong to that current of contemporary political thinking that tries to rehabilitate virtue in the political and social realms, e.g. communitarianism and neo-republicanism. In the words of Robyn Eckersley: "green democratic theory seeks the politicisation of the private good as well as the re-politicisation of the public good" (Eckersley 2004: 96). However, as referred to nature, this politicisation seems to depend on the previous spreading – or imposition? – of ecocentric values. A deep interpretation of ecological citizenship is thus attached to an *ecological identity* that the individual is encouraged to assume, an identity wherefrom the right green behaviour stems. Such an identity is rooted in a moral view of socionatural relationships that leads to a personal assumption of responsibilities that are in turn translated into duties. Yet this ethical deepness overburdens ecological citizenship, linking it to a moral perfectionism that, again, does not seem – if desirable – generalisable. As John Barry puts it, this approach

> demands too much, especially in the absence of any discussion of the balance to be struck between legitimate 'self-interest' and concern for others (..) [it] does not specify the reasons why ecological citizens care; rather it is assumed that by definition they care (Barry 2002: 146).

Furthermore, a thick moralisation of this sort poses insurmountable problems for the institutionalisation of ecological citizenship's content. Moral duties and obligations belong – insofar as they are just that, moral – outside the legal system. Of course, some of these duties can be incorporated into the set of legally binding norms that constitute environmental law, but some of them not. We could even say that the first duty of *any* citizen is to abide by such laws. The rest is virtue.

It is in this context that Derek Bell (2005: 190) has defended the notion of "non-enforceable duties" as a foundation for ecological citizenship: duties that are suggested, but not imposed, by the state. They would be considered as public virtues, an aspect of the identity of a citizen qua citizen (Hailwood 2005: 203).

This makes sense: some citizens are better than others in almost every conceivable realm. Therefore, the actual practice of some green virtues would *describe* a citizen as an ecological citizen. Hence it is the actual practice of virtue that makes the difference. But we should reject the idea that a previous list of acceptable/unacceptable green virtues can be set. There are, there must be, several ways of being an ecological citizen. Otherwise, the citizen would be forced to fit into a given description of the right ecological virtues. As defined by whom? Nobody can claim a monopoly on this definition. That is why I would like to suggest that we should enlarge the potential content of ecological citizenship by connecting it to an open conception of sustainability.

Ecological Citizenship and Open Sustainability

It can then be argued that there are many ways of being green, because there are many (albeit not infinite) ways of conceiving the sustainable society. Some of these ways involve taking care of the remaining natural world, some involve a reasonable use of resources, while still others entail a preference for technological innovation or, why not, the opposite. There is no such thing as the *right* green attitude, since there is no such thing as the *right* sustainable society. The supporter of a particular view of the latter, of course, will contest this claim – all the more if it holds a position that is hardly compatible with alternative approaches to sustainability, as is the case with ecocentrism or any other deep green approach.

Now, in a liberal society that incorporates sustainability as a public goal, the most relevant rules of environmental behaviour for individuals and associations of any kind will be, sooner or later, translated into law. They will be enforceable. If we take for granted that any citizen must abide by any law, what does it take to be an ecological citizen in a free green society? It takes, I suggest, doing *something* on behalf of a particular view of the sustainable society. It must be a free and conscious behaviour, be it a practical or a communicative action. But there is no previous restriction regarding the kind of behaviour that qualifies as acceptable for a citizen to be "ecological". This entails recognition of the complexity of contemporary life and its many opportunities and layers of action:

> In sum, the source of the complexity I attribute to green citizenship is the intensely complex structure of contemporary political life. (...) The notion of complex green citizenship is based on the simple recognition that different kinds of political activities are possible and effective, in a word appropriate, in different contexts: green citizenship is complex precisely (Trachtenberg 2010: 346).

Yet we can also refer to the complexity of contemporary life and to different kinds of activities without adding the adjective "political". It is not necessary to behave *politically* to make a contribution to the preferred sustainable society. In a trivial sense, needless to say, everything is political. But the kind of actions that an ecological citizen can perform needs not to be explicitly political. It can be just a civic, domestic, or economic action – from supporting a green campaign to saving energy to purchasing some certain goods. These actions do have political significance, insofar as they lend support *de facto* to a particular view of sustainability, but are not forcefully political.

I would like to argue that this approach to ecological citizenship can benefit from a number of insights related to the idea of a deep citizenship, i.e. an active conception of citizenship that tries to go beyond its classical understanding as a set of entitlements. This is not an obvious connection, but it is a useful one. The classical roots of a deeper view of citizenship are evident in Hannah Arendt's (1999) conception of the political, according to which the individual is actually born again through participation in the public sphere. Such participation is conceived, in Aristotelian terms, as an end in itself. Of course, there is no *polis* anymore: the political action must be adapted to a different public space. Drawing on Arendt, Sheyla Benhabib (1992: 93) has proposed an *associational* conception of the public space, a space that emerges *whenever and wherever* men act together in concert. Therefore, the public sphere ceases to be a topographical or institutional place, being created instead by the action of citizens themselves. The limits of the political are thus blurred. As Barry Clarke puts it:

> an act of deep citizenship transcends these traditional distinctions. Deep citizenship is concerned less with the domain within which the act takes place than with the trajectory of an act. An act, from whatever domain, public or private, that is oriented to the particular, the private or the sectional is not an act of citizenship. By contrast an act from whatever domain that is oriented towards the universal is an act of deep citizenship (Barry Clarke 1996: 82).

But, as I said, an act of deep citizenship does not have to be directly political. It just requires self-consciousness and a sense of purpose, so that we can distinguish it, as we will see in a moment, from those behaviours that simply follow a cultural or sociological inertia. In that sense they are *creative*, no matter how modest they may appear. After all, we all participate in the process of creating social reality, a process that can be seen as a daily negotiation adopting the form of a spontaneous order: as we all act, we all give shape to the social landscape. Moreover, it is not clear why a deep citizenship act must be *collective*, as Benhabib suggests when she refers to men "acting together in concert". This might well be the case, but might as well not: an individual action should also be accepted as an expression of citizenship. Domestic or civic actions can be both individual and effective, and the same goes, *mutatis mutandis*, with market actions. It is the (desirable) aggregation of virtuous or beneficial actions that makes the (collective) difference.

Therefore, in a society which has adopted an open sustainability as a public value, citizens can contribute to the latter in several ways. Some of them may barely entail *expressing* green values or *omitting* an environmentally harmful behaviour, whereas *participating* in environmental campaigns or social movements would be in the opposite end of personal commitment. This has been also pointed out from the point of view of civic republicanism: social and economic practices are also related to sustainability (Barry 2006). What counts is that a citizen possesses some degree of environmental awareness and hence defends in some way *some* version of sustainability – regardless of the moral judgment that a supporter of some *other* version can hold. Upon these premises, an ecological citizen can be defined as:

> A citizen who, possessing a changeable degree of commitment, which can be expressed in a number of realms (the moral, the domestic, the public) in several ways, makes some kind of personal contribution to the achievement of some form of sustainable society.

This is an open definition of ecological citizenship that corresponds to an open conception of sustainability. The ensuing coexistence between different behaviours and communicative actions, each expressing a particular view of the sustainable society, will take place in the non-institutional as well as in the institutional realms, so that the state will translate into legislation the cultural changes that become consolidated enough (let us remember that the state will impose the environmental standards that add up to a "minimal" sustainability). As Neuteleers has suggested in connection to green lifestyles, they are a sign to policy-makers and they influence other people through citizen-citizen diffusion (Neuteleers 2010: 514). A lifestyle is a way of living; ecological citizenship is a way of behaving that ends up giving shape to certain lifestyles.

Thus there can be no doubt that ecological citizenship does not have to be conceived in a deep moral way in order to possess political significance. Furthermore, I take this to be a more realistic definition of what ecological citizenship *may* be in practice. Let us not forget that a too strict definition of the latter can be impossible to cope with for actual citizens, who are normally full of duties and short of time (see MacGregor 2006). It is also in this way that the deep green focus on guilt and burden can be compensated with some measure of critical reflectiveness, agency, hope, and opportunity (Gabrielson 2008). A more enticing narrative for sustainability can thus emerge.

Helping Citizens to Be Green?

Nevertheless, a reformulation of ecological citizenship along these lines is no guarantee that enough citizens will engage themselves in an environmentally conscious behaviour. This must not be a tragedy, provided that the state is able to set a number of standards that avoid a path of unsustainability. But then again, it is obvious that a society where this kind of citizen abounds will be more easily sustainable and also a richer, more diverse one – especially if we take sustainability as a framework for discussing the way in which we wish to live. A problem is then how these citizens can emerge without significantly violating their freedom to follow their preferred life-plans.

From the abovementioned empirical studies on the current ecological citizen can be deduced that the educative level is a key factor: the more educated the citizen, the more prone to engage in environmentally virtuous practices. That is, admittedly, fairly predictable. Both the social and the individual relationship to the environment are self-reflective relationships that express an increase in the practical and symbolic refinement of human beings. Someday, sustainable behaviours will have been internalised as *habits* by the majority of citizens. But, alas, that has not happened as of yet. A majority of people, even in the supposedly advanced Western societies, lack that degree of self-consciousness, or refuse to exert it. This comes to show the naïvety of the Stuart Mill of *Utilitarianism* (2001), who, claiming that those manners of existence that employs people's higher faculties are preferred by those citizens equally capable of appreciating lower pleasures as well, came to expect that a sufficient advance in the general standards of education would lead a majority of people towards such richer preferences. Of course, there has been a general improvement, but the majority of individuals are neither as cultivated nor as refined as the enlightened philosophers would have expected. If they were, environmentalism would not even exist!

The question follows as to whether they can be *helped* to be better citizens. Or, in other words, whether there is a way to lead citizens in the right direction without damaging their freedom. In the case of ecological citizenship, for instance, once the liberal state adopts sustainability as a goal toward which society is orientated, the latter would benefit from an increase in the number of citizens that become aware of sustainability, without imposing a particular view of it on them. Society would thus be greener without being (not much) less free. Is this possible? Is it acceptable?

A way of doing so has certainly been proposed, under the suggestive label of "libertarian paternalism" (Thaler and Sunstein 2009). According to this, citizens can be *nudged* to take better decisions for both themselves and for society as a whole, in a way that does not violate their freedom to choose. How? By arranging the context of choice in a manner that would steer people in directions that would improve their lives. For instance, more healthy food can be more prominently displayed in schools, so that more of it is actually chosen; and so forth. To count as a nudge, Thaler and Sunstein claim, the intervention must be easy and cheap to avoid, since it cannot be a mandate, nor a ban: "Libertarian paternalism is a relatively weak, soft, and nonintrusive type of paternalism because choices are not blocked, fenced off, or significantly burdened" (Thaler and Sunstein 2009: 6). A dose of paternalism would be acceptable in the face of the fact that most of the people make decisions full of biases and misconceptions – mainly because of inertia and peer influence. This is governance rather than government.

For all the structural problems that affect the state of the environment, nudges are applicable in this realm too. In fact, the very notion of ecological citizenship is a reminder of the influence that the aggregation of individual decisions can exert on the long run, without denying the importance of government intervention. The environment can thus be conceived as "the outcome of a global choice architecture system in which decisions are made by all kinds of actors, from consumers to large companies to governments" (Thaler and Sunstein 2010: 195). Proper incentives and feedbacks can thus help to promote better decisions. Among the former, cap-and-trade systems or green taxes help to re-design markets without imposing a given behaviour, hence fostering innovation and competition; the latter includes stickers that indicate the ecological footprint of a given product or public inventories that report the chemicals that companies release into the environment; finally, smart metres may count as a combination of incentive and feedback that help consumers to use less energy.

Nudges, then, seek to influence behaviour. Supporters of a morally charged conception of ecological citizenship tend to consider that this is too superficial an intervention to be credible as a foundation for deep societal change. Andrew Dobson has suggested that ecological citizenship works at a deeper level than that, by asking people to reflect on the attitudes that inform their behaviour: "In other words, the environmental citizen's behaviour is informed by a systemic understanding of the problems that lead to the perpetuation of injustice in the form of occupancy of unjust amounts of ecological space" (Dobson 2009: 137). But, again, this would be equal to ascribing a fixed content to ecological citizenship, whereas it seems more apt for a free green society to leave its content *relatively* open. Ecological citizenship should not be conceived *in advance* as a set of values and practices designed to realise strong sustainability – but as one of the means that serve to realise the public value of sustainability writ large. Furthermore, it could be said that nudges on behaviour are more acceptable than interventions on attitudes, which properly belong to the realm of individual reflection. And it would be hardly surprising to witness that a change in behaviours ultimately affect attitudes too.

But who nudges the nudger? In other words, who is to decide which kind of decisions should be promoted by arranging choice architecture in a particular manner? How is possible abuse to be avoided? It should not be that complicated. Nudges should reflect the state of the debate around sustainability in each given moment, translating into the realm of incentives and feedback the ongoing consensus around "minimal" sustainability. Nudges are not a substitute for morality, nor should they replace the process of value-formation that takes place spontaneously within society. In addition to that, private companies and associations must respect the law; governmental nudges must respond to the outcome of democratic procedures, which in turn reflect social change; democratic control devices, from public opinion to courts of law, would prevent abuse. The discussion around nudges, in fact, would operate as a discussion on the good life.

In sum, in a liberal society that has assumed a mandate to sustainability, nudges should be considered acceptable. They are a way of promoting a virtuous behaviour, in a realm where "virtue" is easy to define: a behaviour that contributes to some kind of sustainable society. It must be noted, though, that people are free to *not* act virtuously if they so choose. They can be impervious to environmental issues, prefer a fully technological culture, or defend human superiority over the natural world. Yet they will not be allowed to behave in a way that hinders or prevent the achievement of the "minimal" sustainability embodied in environmental laws. As citizens, they must abide by those laws. On the other hand, not everyone among those who "choose" virtue, however, count as ecological citizens. The reason is clear: practices of citizenship require a certain degree of self-consciousness. Those who just respond to the nudge are not promoting a view of the sustainable society, nor practicing virtue in a reflective manner. By making this distinction, we preserve a special status for ecological citizenship – as an institution based on the self-conscious promotion of *a* view of sustainability. However, it is not necessary either to possess a sophisticated position on the issue in order to count as a promoter of a given sustainable society. It suffices that the citizen believes that some kind of societies are better than others and act correspondingly.

Conclusion:
The Future of Environmentalism

It is somewhat shocking that the first signs of environmentalism's normalisation, namely, the questioning of its own foundations once the socionatural relation has been incorporated into the mainstream political agenda, have been interpreted as a symptom of its possible death (Nordhaus and Shellenberg 2007; Cohen 2006). Paradoxically, the advent of a *post-environmentalism* had been heralded long before this alleged death took place (Young 1990). To make things more amusing, a third possibility has recently emerged, namely, that of an environmentalism which is neither death nor re-animated, but actually a *zombie* (Anderson 2010). Of course, all this has partly to do with some kind of inner impulse that makes any given field of thought prone to exhaust its life cycle as soon as possible, so that it seems just a matter of time that crises are diagnosed and ends announced. Likewise, the academic passion for prefixes and labels play a part in it too. Yet there is more than conceptual inertia or intellectual snobbery in the idea that environmentalism has come to an end.

Naturally, that such event may be described just four decades after environmentalism's birth, precisely in the moment of its triumph, is a way of expressing something else. Namely, that the evolution experienced by green thought entails the end of environmentalism as we knew it. This gradual transformation is far from complete and cannot be taken for granted, since a majority within the green movement is bound to resist it – insofar as they understand environmentalism as a radical ideology committed to a radical societal change. Be that as it may, ironically, the debate about the hypothetical death of environmentalism may also be seen as the best proof of its vitality.

Now, although it is possible to make almost scholastic distinctions between the different meanings that this ending may adopt, the semantic field of such a bold statement can be reduced to a couple of them. They are, in any case, related to each other.

In a first sense, environmentalism would end because it would have been *realised* after its main goals had been achieved. From this point of view, the green movement would have succeeded in crucial regards: ecological crisis is now an important issue in the public agenda, nature is a political category in its own right, a debate is open in liberal societies about how to achieve sustainability and how to protect the natural environment reasonably. Green values were marginal and have become mainstream. Therefore, it is suggested, green thinkers have finished their job and should now give way to those actors that translate cultural change into public policy: "their main task was to enlighten people and make them realize that the latest scientific developments point to the environmental damage which is human-made and threatens humanity and the Earth itself: there is no choice but to see green" (Talshir 2004: 10). Environmentalism's death would then be the natural result of its realisation, the move from the exceptional to the normal, its final integration into cultural mainstream.

Yet most green theorists and activists would deny without hesitation the suggestion that environmentalism has actually achieved its goals. The reason is simple indeed: our society has nothing to do with the sustainable society traditionally advocated by the green movement. Whereas the latter can only be understood as the result of implementing a green programme grounded on the moralisation of the socionatural relations, the current understanding of sustainability seems to be a very shallow reflection of such ideal – and therefore a sustainability that is more liberal than green. Thus the end of environmentalism should be understood more rigorously in a second sense: as the end of a particular way of understanding green politics, that which pertains to classical environmentalism and still represents environmentalism writ large:

> Environmentalism as a counter-movement and counter-ideology, as a critique and alternative to capitalism, liberalism, the Enlightenment and anthropocentrism, it would then seem, is at an end (Levy and Wissenburg 2004c: 194).

This end would then be a consequence of the twofold process explored in this book: the critique of environmentalism's traditional foundations and its incipient convergence with liberalism. At the same time, or as a practical corollary of this, liberal democracy is increasingly accepted as the framework wherein sustainability is to be pursued and as the discursive realm wherein the moral green program is to be defended. A correlation can thus be suggested between the political acceptance of democratic pluralism and the re-interpretation of the human-nature relationship in a more realistic fashion – replacing an abstract idealism with a nuanced understanding of the socionatural metabolism. So that:

> the object of environmentalism should be the conflict between different human conceptions of the good and green society, and not the traditional task of adapting human society to the requirements of an independent non-human world (Levy 2004: 49).

Abandoning metaphysics facilitates the rise of a more pragmatic and flexible green politics. And it is this which makes a proper politicisation of the environment possible, instead of a false democratisation that is based on non-debatable naturalistic premises and leads to a non-negotiable model of the sustainable society. As it happens, the moralisation of socionatural relations can be defended in the public sphere and be pursued in the private one, but it is far from constituting the sole prescription of a sustainable society whose shape cannot be anticipated.

Thus a change in environmentalism's *ends* would take place, according to which the former would no longer pursue the overcoming of liberal democracy as a pre-condition for realising a sustainable society grounded on ecocentric or at least non-anthropocentrical ethical foundations. Yet the *means* of green politics would change correspondingly, in the direction of a full acceptance of a pluralistic democracy. It has already been shown that an obvious contradiction has always existed between the substantive goals of environmentalism and democracy's proceduralist nature – or between liberalism's formal neutrality towards different conceptions of the good life and the asserting of those conceptions as legitimate shaping forces of society. This tension is not easily solvable from a theoretical point of view. That is why the green acceptance of democracy must go beyond rhetoric and express itself in a renewal of environmentalism's foundations. Giving up consequentialism leads to a democratic contingency. And these two processes are connected:

> If liberal democracy is the only appropriate mode of articulation for environmental politics, then environmentalists no longer have a *reason* to be (radical) environmentalists (…) as a fundamental challenge to liberal democratic norms is no longer necessary (Humphrey 2004: 115).

In this regard, the gradual improvements attained by liberal democracies in the environmental realm, irrespective of whether they have been achieved thanks to or in spite of the current sociopolitical system itself, have come to show that some form of sustainability can be achieved within the liberal institutional framework. Thus the seeming incompatibility of environmentalism, sustainability and liberalism starts to dissolve. As a result, environmentalism's *political* pacification might be followed by its *theoretical* pacification, so that its participation in the public conversation of liberal societies becomes utterly normalised (Mills and King 2004: 76). However, it is not clear that such pacification can reach the whole green movement, nor that there can emerge a consensus around the new orientation resulting thereof. In fact, the rise of global warming has provoked a resurgence of classical environmentalism and its rationale of doom and radical change, making the prospect of such pacification even more unlikely for the time being.

It should be noted that being faithful to the environmentalism's foundational values entails the refusal of any form of sustainability that may be achieved *within* liberal democracy, insofar as the latter will necessarily diverge from those that can be deduced from an ecocentric worldview. From this point of view, there can simply be no such thing as a liberal sustainability: liberalism cannot possibly be green. In fact, the very renewal advocated here as an immanent development of green political theory contrarily can be seen, by those who disagree with it, as the mere by-product of a liberal *assimilation* of the green ideal. The renewal of environmentalism would then be just a betrayal – carried out in the name of pragmatism.

Normalisation is thus perceived as capitulation. A liberal environmentalism becomes just another lifestyle deprived of any real transformative force. Such neutralisation would lead to a post-ecological politics which deceives itself by thinking that liberal societies have incorporated the environmental agenda and advance towards sustainability. This state of opinion would have even been prepared by a well-organised negationism that, using a conservative anti-environmentalist rhetoric, discredits the green movement and produces the dangerous delusion of a liberal control of the ecological crisis (Blühdorn 2004b: 36; Buell 2003). The historic identity of environmentalism runs the risk of disappearing:

> There is no doubt that environmentalism, understood as radical green politics, does face serious challenges, especially from the dangers attendant upon its insertion as a "normal and mundane" aspect of (bureaucraticized/administrative) liberal democratic politics, aided by a corporate-based anti-environmentalist backlash, and the pervasive dominance of economistic power/knowledge (Barry 2004: 191).

This radical environmentalism, in fact, remains as active as ever, strengthened as it has been with the advent of climate change. Thus we find calls for connecting sustainability and spirituality (Carroll 2004), the suggestion that anarchism should be the foundation for a radical green politics (Carter 1999), the defence of the community as the basic criteria for social organisation in relation to a politics of place (Curry and McGuire 2002), the invocation of either a "new agrarianism" (Freyfogle 2001) or a complete "ruralisation" of society (see Trainer 2010) as ways out of climate change, as well as the ecocentric vindication of a post-liberal green state (Eckersley 2004). There remains then the notion that an *authentic* environmentalism must pose an all-encompassing alternative to liberal-capitalism. Hence the contrast between the "politics of simulation" developed by the former, which is intended to pretend that something is being done about sustainability, and the "authentic politics" embodied by radical environmentalism, committed to the emancipatory ideals of the Enlightenment (Blühdorn 2007: 58). According to this view, the future of environmentalism lies in its past.

It is true that some idiosyncratic green goals can be overlooked by liberal institutions. That is especially the case with those that have to do with the morality of the socionatural relations – like the defence of animal liberation or the protection of nature's intrinsic value. However, in a context of axiological neutrality, some goals are not for the political system to achieve. So Marcel Wissenburg:

> Issues relating to the ideals of the deeper, ecological version of environmentalism require a change of heart that seems too fundamental to be communicable, not to mention ill-suited to be translated into political action (i.e. policy) (Wissenburg 2004: 65).

The communication with the non-human world is an instance at hand. Its *public* arrangement is as difficult as it is inadvisable, whereas its *private* practice is a matter of individual freedom. Radical greens claim that the moralisation of the public sphere should go hand in hand with the politicisation of the private sphere – so that the defence of a green conception of the good in the public debate should be accompanied by its realisation in people's lives. However, a truly democratic environmentalism must accept that there is a permanent conflict between the institutional nature of legal obligations and the voluntary nature of moral duties. It is actually the impossibility that a liberal framework may satisfy the moral claims of radical environmentalism that hinders the abovementioned theoretical pacification of green politics. In fact, the more environmentalists insist on a moral crusade, the less will they be able to influence the ongoing debate about sustainability.

However, the problem does not lie as much in an unfinished process of theoretical renewal, as in the refusal of a great deal of environmentalism to abandon a radical-cum-naturalistic approach. This could even well lead to a split within environmentalism – separating a democratic green politics that tries to realise sustainability in a liberal context from a radical green ideology whose aim is the overcoming of both liberal institutions and cultural anthropocentrism. This would entail an opposition between the old apocalyptic narrative and a new "politics of possibility" that enhances mankind instead of degrading it (Nordhaus and Shellenberg 2007: 150–53). Radical environmentalism can thus become the marginal strain of a wider green politics – or vice versa. It is true that radicalism is a necessary part of any ecosystem of ideas, both within green thought and outside it. Ironically, radicalism is necessary in order to achieve reform. But it can also hinder, or even frustrate, the emergence of a renewed environmentalism. That is a failure that should be avoided. Otherwise environmentalism may experience the saddest death: that of becoming irrelevant.

We are in a transitional moment. The old environmentalism does not die, a new postnatural alternative has not yet been born. Yet a greater awareness of the need to find a new direction for environmentalism is perceptible:

> A middle way is desperately needed, wherein hard-nosed realism about current threats and obstacles to meaningful change meets constructive criticism about current social and political life, and generates creative and far-reaching but viable and efficacious solutions (Vanderheiden 2011: 108–9).

Thus a more practicable alternative, together with a more motivational approach, must replace "apocalyptic negativity" if behavioural change is to succeed (Schmidt 2005; Anderson 2010: 979). An obvious reason for that is that most of the people still want to live well, something that a world based on bioregions does not seem to provide. Rather than "sinks of hope" that mobilise alarmism amid calls to self-sacrifice, "sources of hope" are needed that connect sustainability to more attractive goals (Princen 2010). Both collectively *and* personally, environmentalism cannot suppress things such as individual aspiration, opportunity and choice without losing ground among the public. Instead, it has to build upon them. Whereas sustainability makes change *necessary*, wellbeing makes it *desirable* (Porritt 2005: xv). This demands a reconsideration of the traditional green attitude to economic growth, since growth, albeit one of a corrected character, is indispensable to sustainability: a stagnant society is an absurd dream. In short, sustainability must invoke a conception of the *good* life, instead of being presented as the only way out of a desperate situation. A further reason to abandon the rhetoric of doom is that, as the gloominess of the prognosis is questionable, so is any eco-political theory predicated on it (Haywood 2004: 7). The very idea that socionatural relations stand now in a critical situation can actually be questioned.

This book has advocated an alternative path for environmentalism. It has suggested that only a realistic turn can lead to a successful green politics for the new century. This means accepting the general premise that nature does not exist anymore: we are living after nature and both environmentalism and sustainability must be grounded on that fact. As nature is within society, society is within nature: this is already a postnature age where "neither nature nor humanity has a singular essence or fundamental nature" (Wapner 2010: 33). The outcome of nature-society co-evolution throughout history is the dissolution of nature into human environment. Thus hybridisation and metabolism are the trademarks of socionatural relations. And this is not due to a perverse human will to dominate, but the predictable result of human adaptive behaviour as a species that sets itself apart from nature. It is not deduced from here that everything is artificial now; obviously, it is not. But nature is not independent from us anymore and it is to be understood as a gradable concept rather than as an absolute category. Moreover, although the remaining natural forms that possess a greater symbolic value for us (as reminders of nature's otherness) deserve protection, sustainability is not primarily concerned with the latter. Instead, it is concerned with the achievement of a durable balance between the social and the natural systems. And environmentalism itself is more about sustainability than it is about nature.

Yet the particular form that a sustainable society is to adopt must remain open, so that it stems from the spontaneous development of a liberal and democratic society, instead of being the result of a pre-given blueprint that imposes a single, monist view of sustainability. That is also why a reframed environmentalism ceases to be obsessed with institutionally conscious mechanisms of decision, such as deliberative democracy, and embraces a more thorough view of the ways, formal and informal, planned and spontaneous, that must lead to sustainability – from ecological citizenship to ecological consumption, from information technologies to scientific and technological innovations. A liberal green society is thus to emerge.

Furthermore, the quest for sustainability should be framed as the key element of a reflexive re-organisation of socionatural relations, based on the notion of human refinement. We are not going through any crisis of civilisation, nor are we on the verge of an ecological disaster. Let us leave sentimentalism and hyperboles aside: environmental problems, serious as they are, can be handled by humanity without giving up the best features of a liberal-democratic society. The challenge of sustainability is an opportunity for heading towards a richer, cleaner and more enlightened society. Sustainability, in short, is not about coming back to nature, but about giving shape to a best version of ourselves. This is a new narrative for environmentalism that is also a meaningful narrative for society writ large. We cannot live in Arcadia, and let us not wish to do so either.

Bibliography

Acselrad, H. 1999. Sustainability and Territory: Meaningful Practices and Material Transformations, in *Sustainability and the Social Sciences. A Cross-disciplinary Approach to Integrating Environmental Considerations into Theoretical Reorientation*, edited by E. Becker and T. Jahn. London: Zed Books, 37–58.

Adger, N. 2007. Ecological and Social Resilience, in *Handbook of Sustainable Development*, edited by G. Atkinson, S. Dietz and E. Neumayer. Cheltenham and Northampton: Edward Elgar, 78–90.

Adger, N. and Jordan, A. 2009. Sustainability: Exploring the Processes and Outcomes of Governance, in *Governing Sustainability*, edited by N. Adger and A. Jordan. Cambridge: Cambridge University Press, 3–31.

Adorno, T.W. 1997. *Aesthetic Theory*. Minnesota: University of Minnesota Press.

Adorno, T.W. 2001. *Minima Moralia: Reflexionen aus dem Beschädigten Leben*. Frankfurt: Suhrkamp.

Adorno, T.W. and Horkheimer, M. 1997. *Dialectic of Enlightenment*. London: Verso.

Agamben, G. 2004. *The Open: Man and Animal*. Stanford: Stanford University Press.

Albert, M. 2006. *Realizing Hope: Life beyond Capitalism*. London: Zed Books.

Anderson, J. 2010. From 'Zombies' to 'Coyotes': Environmentalism Where We Are. *Environmental Politics* 19(6), 973–91.

Anderson, T.L. (ed.) 2004. *You Have to Admit it's Getting Better. From Economic Prosperity to Environmental Quality*. Stanford: Hoover Institution Press.

Arendt, H. 1999. *The Human Condition*. Chicago: University of Chicago Press.

Arias-Maldonado, M. 2007. An Imaginary Solution? The Green Defence of Deliberative Democracy. *Environmental Values* 16(2), 233–52.

Atkinson, G., Dietz, S. and Neumayer, E. 2007. *Handbook of Sustainable Development*. Cheltenham and Northampton: Edward Elgar.

Attfield, R. 1994. *Environmental Philosophy: Principles and Prospects*. Aldershot: Ashgate.

Ayres, R.U. 2000. Commentary on the Utility of the Ecological Footprint Concept. *Ecological Economics* 32, 347–9.

Baber, W.F. and Bartlett, R.V. 2005. *Deliberative Environmental Politics. Democracy and Ecological Rationality*. Cambridge and London: MIT Press.

Bäckstrand, K. 2004. Precaution, Scientization or Deliberation? Prospects for Greening y Democratizing Science, in *Liberal Democracy and Environmentalism. The End of Environmentalism?*, edited by Y. Levy and M. Wissenburg. London: Routledge, 100–112.

Bacon, F. 2010. *Novum Organum*. Forgotten Books.

Baker, S. 2006. *Sustainable Development*. London and New York: Routledge.

Ball, T. 2006. Democracy, in *Political Theory and the Ecological Challenge*, edited by A. Dobson and R. Eckersley. Cambridge: Cambridge University Press, 131–47.

Barber, B. 2004. *Strong Democracy: Participatory Politics for a New Age*. Berkeley and Los Angeles: University of California Press.

Barkow, J., Cosmides, L. and Tooby, J. 1992. *The Adapted Mind: Evolutionary Psychology and the Generation of Culture*. New York: Oxford University Press.

Barry, B. 1979. Is Democracy Special?, in *Philosophy, Politics and Society. Fifth Series*, edited by P. Laslett and W.G. Runciman. Oxford: Basil Blackwell, 155–96.

Barry, J. 1999. *Rethinking Green Politics. Nature, Virtue and Progress*. London: Sage.

Barry, J. 2001. Greening Liberal Democracy: Practice, Theory and Political Economy, in *Sustaining Liberal Democracy. Ecological Challenges and Opportunities*, edited by J. Barry and M. Wissenburg. Houndmills: Palgrave, 59–80.

Barry, J. 2002. Vulnerability and Virtue: Democracy, Dependency and Ecological Stewardship, in *Democracy and the Claims of Nature*, edited by B. Minteer and B. Pepperman Taylor. New York: Rowan & Littlefield, 133–52.

Barry, J. 2004. From Environmental Politics to the Politics of the Environment: The Pacification and Normalization of the Environment?, in *Liberal Democracy and Environmentalism. The End of Environmentalism?*, edited by Y. Levy and M. Wissenburg. London: Routledge, 179–92.

Barry, J. 2006. Resistance is Fertile: From Environmental to Sustainability Citizenship, in *Environmental Citizenship*, edited by A. Dobson and D. Bell. Cambridge and London: The MIT Press, 21–48.

Barry, J. and Wissenburg, M. (eds) 2001. *Sustaining Liberal Democracy. Ecological Challenges and Opportunities*. Houndmills: Palgrave.

Barry, J. and Eckersley, R. (eds) 2005. *The State and the Global Ecological Crisis*. Cambridge: The MIT Press.

Barry Clarke, P. 1996. *Deep Citizenship*. London: Pluto Press.

Barthes, R. 1993. *Mythologies*. London: Vintage Classics.

Basl, J. 2010. Restitutive Restoration: New Motivations for Ecological Restoration. *Environmental Ethics* 32(2), 135–47.

Bataille, G. 1989. *Theory of Religion*. New York: Zone Books.

Baxter, B. 1999. *Ecologism. An Introduction*. Edinburgh: Edinburgh University Press.

Baxter, B. 2007. *A Darwinian Worldview. Sociobiology, Environmental Ethics and the Work of Edward O. Wilson*. Aldershot: Ashgate.

Bayertz, K. 2005. Die Menschliche Natur und ihr Moralischer Status, in *Die Menschliche Natur. Welchen und wiviel Wert hat sie?*, edited by K. Bayertz. Paderborn: Mentis, 9–31.

Bealey, F. 1988. *Democracy in the Contemporary State*. Oxford: Clarendon Press.

Beck, U. 1992. *Risk Society. Towards a New Modernity*. London: Sage.

Beckerman, W. 1994. Sustainable Development: Is it a Useful Concept? *Environmental Values* 3(3), 191–209.

Beckerman, W. 2002. *A Poverty of Reason. Sustainable Development*. Oakland: The Independent Institute.

Beckman, L. 2001. Virtue, Sustainability and Liberal Values, in *Sustaining Liberal Democracy. Ecological Challenges and Opportunities*, edited by J. Barry and M. Wissenburg. Houndmills: Palgrave, 179–91.

Bell, D. 2005. Liberal Environmental Citizenship. *Environmental Politics* 14(2), 179–94.

Benhabib, S. 1992. *Situating the Self*. Cambridge: Polity.

Benhabib, S. 1996. Toward a Deliberative Model of Democratic Legitimacy, in *Democracy and Difference. Contesting the Boundaries of the Political*, edited by S. Benhabib. Princeton: Princeton University Press, 67–94.

Benjamin, W. 2009. *One-way Street and Other Writings*. London: Penguin.

Benton, T. 1992. Ecology, Socialism and the Mastery of Nature: A Reply to Reiner Grundmann. *New Left Review* 194, 55–74.

Benton, T. 1993a. *Natural Relations. Ecology, Animal Rights and Social Justice*. London: Verso.

Benton. T. 1993b. Animal Rights and Social Relations, in *The Politics of Nature*, edited by A. Dobson and P. Lucardie. London: Routledge, 161–76.

Benton. T. 1994. Biology and Social Theory in the Environmental Debate, in *Social Theory and the Global Environment*, edited by M. Redclift and T. Benton. London: Routledge, 28–50.

Berger, P. and Luckmann, T. 1991. *The Social Construction of Reality: A Treatise in the Sociology of Knowledge*. London: Penguin.

Biesecker, A. and Hofmeister, S. 2006. *Die Neuerfindung des Ökonomischen. Ein (re)produktionstheoretischer Beitrag zur Sozialökolosgischen Forschung*. Munich: Oekom.

Biesecker, A. and Hofmeister, S. 2009. Starke Nachhaltigkeit fordert eine Ökonomie der (Re)Produktivität, in *Die Greifswalder Theorie starker Nachhaltigkeit*, edited by T. Egan-Krieger et al. Marburg: Metropolis-Verlag, 169–92.

Birnbacher, D. 2006. *Natürlichkeit*. Berlin: Walter de Gruyter.

Black, J. 1970. *The Dominion of Man. The Search for Ecological Responsibility*. Edinburgh: Edinburgh University Press.

Blühdorn, I. 2000. Ecological Modernisation and Post-ecological Politics, in *Environment and Global Modernity*, edited by G. Spaargaren and F. Buttel. London: Sage, 209–28.

Blühdorn, I. 2004a. *Post-ecologist Politics: Social Theory and the Abdication of the Ecologist Paradigm*. London: Routledge.

Blühdorn, I. 2004b. Post-ecologism and the Politics of Simulation, in *Liberal Democracy and Environmentalism. The End of Environmentalism?*, edited by Y. Levy and M. Wissenburg. London: Routledge, 35–47.

Blühdorn, I. 2007. Sustaining the Unsustainable: Symbolic Politics and the Politics of Simulation. *Environmental Politics* 16(2), 251–75.

Blühdorn, I. and Welsh, I. 2007. Eco-politics beyond the Paradigm of Sustainability: A Conceptual Framework and Research Agenda. *Environmental Politics* 16(2), 185–205.

Boaz, D. 1998. *Libertarianism*. New York: The Free Press.

Bobbio, N. 1990. *Liberalism and Democracy*. London: Verso.

Bohman, J. and Rehg, W. 1997. Introduction, in *Deliberative Democracy. Essays on Reason and Politics*, edited by J. Bohmand and W. Rehg. Cambridge: The MIT Press, ix–xxx.

Bookchin, M. 1980. *Toward an Ecological Society*. Montreal: Black Rose Books.

Bookchin, M. 1986. *The Modern Crisis*. Philadelphia: New Society Publishers.

Bookchin, M. 2001. What is Social Ecology?, in *Environmental Ethics*, edited by M. Boylan. Upper Saddle River: Prentice Hall.

Bosselmann, K. 2008. *The Principle of Sustainability*. Aldershot: Ashgate.

Botsman, R. and Rogers, R. 2011. *What's Mine is Yours: How Collaborative Consumption is Changing the Way We Live*. New York: HarperCollins.

Brand, S. 2009. *Whole Earth Discipline*. London: Atlantic Books.

Brechin, S. 2010. Public Opinion: A Cross-national View, in *Routledge Handbook of Climate Change and Society*, edited by C. Lever-Tracy. London: Routledge, 179–209.

Brennan, A. and Lo, Y.S. 2010. *Understanding Environmental Philosophy*. Durham: Acumen.

Buell, F. 2003. *From Apocalypse to Way of Life: Environmental Crisis in the American Century*. New York: Routledge.

Bukharin, N. 1972. *Teoría del materialismo histórico*. Madrid: Siglo XXI. (English edition: *Historic Materialism*. Michigan: The University of Michigan Press, 1970.)

Burnigham, K. and Cooper, G. 1999. Being Constructive: Social Constructionism and the Environment. *Sociology* 33(2), 297–316.

Caes, C.J. 1995. *Natural U. Learning from Nature*. Lanham: University Press of America.

Cafaro, P. and Staples, W. 2009. The Environmental Argument for Reducing Inmigration into the United States. *Environmental Ethics* 31, 3–28.

Capra, F. 1996. *Web of Life: A New Synthesis of Mind and Matter*. New York: Anchor Books.

Carroll, J. 2004. *Sustainability and Spirituality*. Albany: State University of Nueva York Press.

Carter, A. 1999. *A Radical Green Political Theory*. London: Routledge.

Carter, A. 2001. *The Political Theory of Global Citizenship*. London: Routledge.

Carter, N. 2007. *The Politics of the Environment*, 2nd edition. Cambridge: Cambridge University Press.

Cassirer, E. 1972. *An Essay on Man. An Introduction to a Philosophy of Human Culture*. New Haven and London: Yale University Press.

Castells, M. 2009. *Communication Power*. Oxford: Oxford University Press.

Castro Nogueira, L., Castro Nogueira, L. and Castro Nogueira, M.A. 2008. *¿Quién teme a la naturaleza humana?* Madrid: Tecnos.

Cavallisforza, L.L. and Feldman, M.W. 1981. *Cultural Transmission and Evolution: A Quantitative Approach*. Princeton: Princeton University Press.

Chiesura, A. and de Groot, R. 2003. Critical Natural Capital: A Socio-cultural Perspective. *Ecological Economics* 44, 219–31.

Choucri, N. 1999. The Political Logic of Sustainability, in *Sustainability and the Social Sciences. A Cross-disciplinary Approach to Integrating Environmental Considerations into Theoretical Reorientation*, edited by E. Becker and T. Jahn. London: Zed Books, 143–61.

Cleveland, C.J., Hall, C.A.S. and Kaufmann, R. 1984. Energy and the US Economy: A Biophysical Perspective. *Science* 255, 890–97.

Cocks, D. 2003. *Deep Futures. Our Prospects for Survival*. Montreal: University of New South Wales Press.

Cohen, J. 1998. Democracy and Liberty, in *Deliberative Democracy*, edited by J. Elster. Cambridge: Cambridge University Press, 185–231.

Cohen, M. 2006. The Death of Environmentalism. *Organization & Environment* 19(1), 74–81.

Commoner, B. 1971. *The Closing Circle. Confronting the Environmental Crisis*. London: Jonathan Cape.

Cooper, D.E. 1998. Aestheticism and Environmentalism, in *Spirit of the Environment. Religion, Value and Environmental Concern*, edited by D. Cooper and J.C. Palmer. London: Routledge, 100–112.

Costanza, R. 2000. The Dynamics of the Ecological Footprint Concept. *Ecological Economics* 32, 341–5.

Costanza, R. and Daly, H.E. 1992. Natural Capital and Sustainable Development. *Journal of Conservation Biology* 6, 37–46.

Cotgrove, Stephen. 1982. *Catastrophe of Cornucopia. The Environment, Politics and the Future*. Chichester: John Wiley and Sons.

Cronon, W. 1996. *Uncommon Ground. Rethinking the Human Place in Nature*. New York: W.W. Norton & Company.

Crutzen, P.J. and Stoermer, E.F. 2000. The Anthropocene. *Global Change Newsletter* 41, 17–18.

Curry, J. and McGuire, S. 2002. *Community on Land. Community, Ecology and the Public Interest*. Lanham: Rowman & Littlefield.

Dahl, R. 1979. Procedural Democracy, in *Philosophy, Politics and Society. Fifth Series*, edited by P. Laslett and W.G. Runciman. Oxford: Basil Blackwell, 97–133.

Darwin, C. 2008. *On the Origin of Species*. Oxford: Oxford University Press.

Davidson, J. 2000. Sustainable Development: Business as Usual or a New Way of Living? *Environmental Ethics* 22, 25–42.

Dawson, A. 1999. The Problem of Pigs, in *Geography and Ethics. Journeys in a Moral Terrain*, edited by J.D. Proctor and D.M. Smith. London: Routledge, 193–205.

Deacon, T. 1997. *The Symbolic Species. The Co-evolution of Language and the Human Brain*. London: Penguin.

De Geus, M. 1999. *Ecological Utopias. Envisioning the Sustainable Society*. Utrecht: International Books.

De Groot, R.S. 1992. *Functions of Nature*. Amsterdam: Wolters-Noordhoff.

Derber, C. 2010. *Greed to Green. Solving Climate Change and Remaking the Economy*. Boulder & London: Paradigm Publishers.

Derrida, J. 2008. *The Animal that Therefore I Am*. New York: Fordham University Press.

Dessler, A. and Parson, E. 2006. *The Science and Politics of Global Climate Change. A Guide to the Debate*. Cambridge: Cambridge University Press.

Deutsch, L., Folke, C. and Skanberg, K. 2003. The Critical Natural Capital of Ecosystem Performance as Insurance for Human Well-being. *Ecological Economics* 44, 205–17.

De-Shalit, A. 1995. *Why Posterity Matters. Environmental Policies and Future Generations*. London: Routledge.

De-Shalit, A. 2000. *The Environment: Between Theory and Practice*. Oxford: Oxford University Press.

Dessler, A.E. and Parson, E.A. 2006. *The Science and Politics of Global Climate Change. A Guide to the Debate*. Cambridge: Cambridge University Press.

Devall, B. and Sessions, G. 1985. *Deep Ecology. Living as if Nature Mattered*. Salt Lake City: Gibbs Smith.

Diamond, J. 2006. *Collapse: How Societies Choose to Fail or Survive*. London: Penguin.

Dickens, P. 1992. *Society and Nature. Towards a Green Social Theory*. London: Harvester Wheatsheaf.

Dickens, P. 1996. *Reconstructing Nature. Alienation, Emancipation and the Division of Labour*. London: Routledge.

Dietz, S. and Neumayer, E. 2009. Economics and the Governance of Sustainable Development, in *Governing Sustainability*, edited by W.N. Adler and A. Jordan. Cambridge: Cambridge University Press, 259–82.

Dobson, A. 1990. *Green Political Thought*, 1st edition. Oxford: Oxford University Press.

Dobson, A. 1996. Democratising Green Theory: Preconditions and Principles, in *Democracy and Green Political Thought*, edited by B. Doherty and M. de Geus. London: Routledge, 132–48.

Dobson, A. 1998. *Justice and the Environment. Conceptions of Environmental Sustainability and Theories of Distributive Justice*. Oxford: Oxford University Press.

Dobson, A. (ed.) 1999. *Fairness and Futurity. Essays on Environmental Sustainability and Social Justice*. Oxford: Oxford University Press.

Dobson, A. 2000a. *Green Political Thought*, 3rd edition. Oxford: Oxford University Press.

Dobson, A. 2000b. Ecological Citizenship: A Disruptive Influence?, in *Politics at the Edge*, edited by C. Pierson. London: Macmillan.

Dobson, A. 2003. *Citizenship and the Environment*. Oxford: Oxford University Press.

Dobson, A. 2007. *Green Political Thought*, 4th edition. Oxford: Oxford University Press.

Dobson, A. 2009. Citizens, Citizenship and Governance for Sustainability, in *Governing Sustainability*, edited by N. Adger and A. Jordan. Cambridge: Cambridge University Press, 125–41.

Dobson, A. 2010. Democracy and Nature: Speaking and Listening. *Political Studies* 58(4), 752–68.

Dobson, A. and Bell, D. (eds) 2006. *Environmental Citizenship*. Cambridge and London: The MIT Press.

Dobson, A. and Lucardie, P. (eds) 1993. *The Politics of Nature*. London: Routledge, 229–34.

Dobson, A. and Valencia Sáiz, A. (eds) 2006. *Citizenship, Environment, Democracy*. London: Routledge.

Doherty, B. and De Geus, M. (eds) 1996. *Democracy and Green Political Thought*. London: Routledge.

Döring, R. 2009. Natural Capital – What's the Difference?, in *Sustainability, Natural Capital and Nature Conservation*, edited by R. Döring. Marburg: Metropolis-Verlag, 123–42.

Downing, L. and Thigpen, R. 1989. A Defense of Neutrality in Liberal Political Theory, *Polity* XXI(3), 502–16.

Drenthen, M. 2009. Ecological Restoration and Place Attachment: Emplacing Non-Places? *Environmental Values* 18, 285–312

Drummond, I. and Marsden, T. 1999. *The Condition of Sustainability*. London: Routledge.

Dryzek, J. 1990. *Discursive Democracy*. Cambridge: Cambridge University Press.

Dryzek, J. 1994. Ecology and Discursive Democracy: Beyond Liberal Democracy and the Administrative State, in *The Politics of the Environment*, edited by R. Goodin. Aldershot: Edward Elgar, 394–418.

Dryzek, J. 1995. Political and Ecological Communication. *Environmental Politics* 4(4), 13–30.

Dryzek, J. 1996. *Democracy in Capitalist Times. Ideals, Limits and Struggles.* New York: Oxford University Press.

Dryzek, J. 2000. *Deliberative Democracy and Beyond. Liberals, Critics, Contestations.* Oxford: Oxford University Press.

Dryzek, J. 2005. *The Politics of the Earth: Environmental Discourses.* Oxford: Oxford University Press.

Dryzek, J. 2006. *Deliberative Global Politics: Discourse and Democracy in a Divided World.* Cambridge: Polity.

Duncan, E. 2009. Getting Warmer: A Special Report on Climate Change and the Carbon Economy. *The Economist*, 5 December.

Dunlap, T.R. 2004. *Faith in Nature. Environmentalism as Religious Quest.* Seattle and London: University of Washington Press.

Dunlap, T.R. and Catton, W. 1994. Struggling with Human Exemptionalism: The Rise, Decline and Revitalization of Environmental Sociology. *American Sociologist* Spring, 5–30.

Dyer, G. 2008. *Climate Wars.* Melbourne: Scribe.

Eckersley, R. 1992. *Environmentalism and Political Theory.* New York: State University of New York Press.

Eckersley, R. 1996. Greening Liberal Democracy. The Rights Discourse Revisited, in *Democracy and Green Political Thought*, edited by B. Doherty and M. de Geus. London: Routledge, 212–36.

Eckersley, R. 1998. Environment Rights and Democracy, in *Political Ecology. Local and Global*, edited by D. Bell et al. London: Routledge, 353–76.

Eckersley, R. 2000. Deliberative Democracy, Ecological Representation and Risk: Towards a Democracy of the Affected, in *Democratic Innovation. Deliberation, Representation y Association*, edited by M. Saward. London: Routledge, 117–32.

Eckersley, R. 2002. Environmental Pragmatism, Ecocentrism, and Deliberative Democracy. Between Problem-Solving and Fundamental Critique, in *Democracy and the Claims of Nature*, edited by B. Minteer and B. Pepperman Taylor. New York: Rowan & Littlefield, 49–69.

Eckersley, R. 2004. *The Green State.* Cambridge: The MIT Press.

Eder, K. 1995. Rationality in Environmental Discourse: A Cultural Approach, in *Green Politics Three*, edited by W. Rüdig. Edinburgh: Edinburgh University Press, 9–37.

Eder, K. 1996. *The Social Construction of Nature.* London: Sage.

Eder, K. 2000. Taming Risks through Dialogues: The Rationality and Functionality of Discursive Institutions in Risk Society, in *Risk in the Modern Age. Social Theory, Science and Environmental Decision-Making*, edited by M. Cohen. London: Macmillan, 225–48.

Edwards, P.N. 2010. *A Vast Machine. Computers Models, Climate Data, and the Politics of Global Warming.* Cambridge and London: The MIT Press.

Ehrlich, P. 1969. *The Population Bomb.* New York: Sierra Club.

Ekins, P. 2000. *Economic Growth and Environmental Sustainability. The Prospects for Green Growth.* London: Routledge.

Ekins, P. et al. 2003. A Framework for the Practical Application of the Concepts of Critical Natural Capital and Strong Sustainability. *Ecological Economics* 44, 165–85.

Elden, S. 2006. Heidegger's Animals. *Continental Philosophy Review* 39, 273–91.

Elliot, K. 2007. Norton's Conception of Sustainability: Political, Not Metaphysical. *Environmental Ethics* 29, 3–22.

Elliot, R. 1982. Faking Nature. *Inquiry* 25, 81–93.

Elliot, R. 1997. *Faking Nature. The Ethics of Ecological Restoration.* London: Routledge.

Elster, J. 1997. The Market and the Forum: Three Varieties of Political Theory, in *Deliberative Democracy. Essays on Reason and Politics*, edited by J. Bohman and W. Rehg. Cambridge: The MIT Press, 3–34.

Elster, J. 1998. Deliberation and Constitution Making, in *Deliberative Democracy*, edited by J. Elster. Cambridge: Cambridge University Press, 97–122.

Escobar, O. 2009. The Dialogic Turn: Dialogue for Deliberation. *In-Spire* 4(2), 42–70.

Estlund, D. 2008. *Democratic Authority. A Philosophical Framework.* Princeton and Oxford: Princeton University Press.

Evernden, N. 1992. *The Social Creation of Nature.* Baltimore: The Johns Hopkins University Press.

Faber, M. and Manstetten, R. 2003. *Mensch-Natur-Wissen. Grundlagen der Umweltbildung.* Gottingen: Vandenhoeck & Ruprecht.

Faber, M., Proops, J. and Manstetten, R. 1997. *Evolution, Time, Production and the Environment.* New York: Springer.

Fischer-Kowalski, M. and Haberl, H. 2007. Conceptualizing, Observing and Comparing Socioecological Transitions, in *Socioecological Transitions and Global Change. Trajectories of Social Metabolism and Land Use*, edited by M. Fischer-Kowalski and H. Haberl. Cheltenham: Edward Elgar, 1–30.

Fischer-Kowalski, M. and Weisz, H. 1999. Society as Hybrid between Material and Symbolic Realms. Toward a Theoretical Framework of Society–Nature Interaction. *Advances in Human Ecology* 8, 215–51.

Foster, J. 2005a. Making Sense of Stewardship: Metaphorical Thinking and the Environment, in *Learning, Natural Capital and Sustainable Development. Options for an Uncertain World*, edited by J. Foster and S. Gough. Abingdon: Routledge, 21–32.

Foster, J. 2005b. Options, Sustainability Policy and the Spontaneous Order, in *Learning, Natural Capital and Sustainable Development. Options for an Uncertain World*, edited by J. Foster and S. Gough. Abingdon: Routledge, 111–31.

Foster, J. 2008. *The Sustainability Mirage. Illusion and Reality in the Coming War on Climate Change.* London: Earthscan.

Foster, J. and Gough, S. (eds) 2005. *Learning, Natural Capital and Sustainable Development. Options for an Uncertain World*. Abingdon: Routledge.

Fraser, N. and Gordon, L. 1994. Civil Citizenship against Social Citizenship?, in *The Condition of Citizenship*, edited by B. van Steenbergen. London: Sage, 90–107.

Freyfogle, E. (ed.) 2001. *The New Agrarianism. Land, Culture and the Community of Life*. Washinton, DC: Island Press.

Friedman, T. 2009. Our One-party Democracy. *New York Times*, 8th September, 29.

Funtowicz, S. and Ravetz, J. 1993. Science for the Post-normal Age. *Futures*. September, 739–55.

Gabrielson, T. 2008. Green Citizenship: A Review and Critique. *Citizenship Studies* 12(4), 429–46.

Gallopín, G. and Raskin, P. 2002. *Global Sustainability. Bending the Curve*. London: Routledge.

Gastil, J. 2008. *Political Communication and Deliberation*. Thousand Oaks/London: Sage.

Gastil, J. and Levine, P. 2005. *The Deliberative Democracy Handbook: Strategies for Effective Civic Engagement in the Twenty-first Century*. San Francisco: John Wiley and Sons.

Georgescu-Roegen, N. 1971. *The Entropy Law and Economic Process*. Cambridge: Harvard University Press.

Giddens, A. 1991. *The Consequences of Modernity*. Cambridge: Polity.

Glacken, C. 1967. *Traces on the Rodian Shore: Nature and Culture in Western Thought from Ancient Times to the End of the Eighteenth Century*. Berkeley: University of California Press.

Glover, L. 2006. *Postmodern Climate Change*. London: Routledge.

Godelier, M. 1986. *The Mental and the Material: Thought, Economy and Society*. London: Blackwell Verso.

Goldsmith, E. 1974. *Blueprint for Survival*. New York: New American Library.

Goodin, R. 1992. *Green Political Theory*. London: Polity.

Goodin, R. 1996. Enfranchising the Earth, and its Alternatives. *Political Studies* XLIV, 835–49.

Goodin, R. 2003. *Reflective Democracy*. Oxford and New York: Oxford University Press.

Görg, C. 2004. The Construction of Societal Relationships with Nature. *Poiesis Prax* 3, 22–36.

Greer, J.M. 2008. *Long Descent: A User's Guide to the End of the Industrial Age*. Gabriola Island: New Society Publishers.

Grober, U. 2010. *Die Entdeckung der Nachhaltigkeit. Kulturgeschichte eines begriffs*. Munich: Antje Kunstmann.

Grundmann, R. 1991a. The Ecological Challenge to Marxism. *New Left Review* 187, 103–20.

Grundmann, R. 1991b. *Marxism and Ecology*. Oxford: Clarendon Press.

Grunwald, A. 2009. Konzepte nachhaltiger Entwicklung vergleichen – aber wie? Diskursebenen und Vergleichsmaßstäbe, in *Die Greifswalder Theorie starker Nachhaltigkeit*, edited by T. Egan-Krieger et al. Marburg: Metropolis-Verlag, 41–64.

Gutmann, A. and Thompson, D. 1996. *Democracy and Disagreement*. Cambridge: Harvard University Press.

Gutmann, A. and Thompson, D. 2004. *Why Deliberative Democracy?* Princeton: Princeton University Press.

Habermas, J. 1996. Three Normative Models of Democracy, in *Democracy and Difference. Contesting the Boundaries of the Political*, edited by S. Benhabib. Princeton: Princeton University Press, 21–30.

Habermas, J. 1997. *Between Facts and Norms: Contributions to a Discourse Theory of Law and Democracy*. Cambridge: Polity.

Habermas, J. 2005. *Die Zukunft der menschlichen Natur. Auf dem Weg zu einer liberalen Eugenik?* Frankfurt: Suhrkamp.

Habermas, J. 2008. *Ach, Europa*. Frankfurt: Suhrkamp.

Haila, Y. and Dyke, C. (eds) 2006. *How Nature Speaks. The Dynamics of the Human Ecological Condition*. Duke: Duke University Press.

Hailwood, S. 2004. *How to Be a Green Liberal: Nature, Value and Liberal Philosophy*. Chesham: Acumen.

Hailwood, S. 2005. Environmental Citizenship as Reasonable Citizenship. *Environmental Politics* 14(2), 195–210.

Haraway, Donna. 2007. *When Species Meet*. Minnesota: University of Minnesota Press.

Hardin, G. 1977a. The Tragedy of the Commons, in *Managing the Commons*, edited by G. Hardin and J. Baden. San Francisco: W.H. Freeman and Company, 16–30.

Hardin, G. 1977b. Living on a Lifeboat, in *Managing the Commons*, edited by G. Hardin and J. Baden. San Francisco: W.H. Freeman and Company, 261–79.

Hardin, R. 1999. Deliberation: Method, Not Theory, in *Deliberative Politics. Essays on Democracy and Disagreement*, edited by S. Macedo. Oxford: Oxford University Press, 103–20.

Hards, S. 2011. Social Practice and the Evolution of Personal Environmental Values. *Environmental Values* 20, 23–42.

Harrison, N. 2000. *Constructing Sustainable Development*. New York: State University of New York Press.

Harrison, R. 1998. Toward a Philosophy of Nature, in *Uncommon Ground. Rethinking the Human Place in Nature*, edited by W. Cronon. New York: W.W. Norton & Company, 426–37.

Harvey, D. 1996. *Justice, Nature & the Geography of Difference*. London: Blackwell.

Hayek, F. 2008. *The Constitution of Liberty*. London: Routledge.

Hayward, T. 1992. *Ecology and Human Emancipation, Radical Philosophy* 62, 3–13.

Hayward, T. 1995. *Ecological Thought: An Introduction*. London: Polity.

Hayward, T. 1998. *Political Theory and Ecological Values*. Cambridge: Polity.

Hediger, W. 2009. The Conceptual Strength of Weak Sustainability, in *Sustainability, Natural Capital and Nature Conservation*, edited by R. Döring. Marburg: Metropolis-Verlag, 21–48.

Heidegger, M. 1995. *The Fundamental Concepts of Metaphysics. World, Finitude, Solitude*. Bloomington and Indianapolis: Indiana University Press.

Heilbroner, R. 1975. *An Inquiry into the Human Prospect*. London: Calder & Boyars.

Heise, U. 2010. *Nach der Natur. Das Artensterben und die Moderne Kultur*. Berlin: Suhrkamp Verlag.

Hettinger, N. 2005. Respecting Nature's Autonomy in Relationship with Humanity, in *Recognizing the Autonomy of Nature. Theory and Practice*, edited by T. Heyd. New York: Columbia University Press, 86–98.

Hinchman, L. 2004. Is Environmentalism a Humanism? *Environmental Values* 13(1), 3–29.

Holland, A. 1994. Natural Capital, in *Philosophy and the Natural Environment*, edited by R. Atfield and A. Belsey. Cambridge: Cambridge University Press, 169–82.

Holland, A. 1999. Sustainability: Should We Start from Here?, in *Fairness and Futurity. Essays on Environmental Sustainability and Social Justice*, edited by A. Dobson. Oxford: Oxford University Press, 46–68.

Hourdequin, M. 2010. Climate, Collective Action and Individual Ethical Obligations. *Environmental Values* 19, 443–64.

Hulme, M. 2009. *Why We Disagree About Climate Change. Understanding Controversy, Inaction and Opportunity*. Cambridge: Cambridge University Press.

Hulme, M. and Neufeldt, H. 2010. Preface: The ADAM Project, in *Making Climate Change Work for Us: European Perspectives on Adaptation and Mitigation Strategies*, edited by M. Hulme and H. Neufeldt. Cambridge: Cambridge University Press, xix–xxv.

Humphrey, M. 2002. *Preservation Versus the People? Nature, Humanity and Political Philosophy*. Oxford: Oxford University Press.

Humphrey, M. (ed.) 2003. *Political Theory and the Environment: A Reassessment*. London: Frank Cass.

Humphrey, M. 2004. Ecology, Democracy and Autonomy: A Problem of Wishful Thinking, in *Liberal Democracy and Environmentalism. The End of Environmentalism?*, edited by Y. Levy and M. Wissenburg. London: Routledge, 115–26.

Humphrey, M. 2008. Seeing is Believing? Aesthetics and the Politics of the Environment. *Environmental Politics* 17(1), 138–46.

Irwin, A. 2001. *Sociology and the Environment*. Cambridge: Polity.

Jackson, R. 2002. *The Earth Remains Forever. Generations at a Crossroads*. Austin: University of Texas Press.

Jackson, T. 2007. Sustainable Consumption, in *Handbook of Sustainable Development*, edited by G. Atkinson et al. Cheltenham and Northampton, 254–68.

Jackson, T. 2009. *Prosperity without Growth: Economics for a Finite Planet*. London: Earthscan.

Jacobs, M. 1997. Environmental Valuation, Deliberative Democracy and Public Decision-Making Institutions, in *Valuing Nature? Ethics, Economics and the Environment*, edited by J. Foster. London: Routledge, 211–31.

Jacobs, M. 1999. Sustainable Development as a Contested Concept, in *Fairness and Futurity. Essays on Environmental Sustainability and Social Justice*, edited by A. Dobson. Oxford: Oxford University Press, 21–45.

Jagers, S. 2009. In Search of the Ecological Citizen. *Environmental Politics* 18(1), 18–36.

Janich, P. 2010. *Der Mensch und andere Tiere. Das zweideutige Erbe Darwins*. Berlin: Suhrkamp.

Jasanoff, S. 2007. Technologies of Humility. *Nature* 450, 33–5.

Jayal, N. 2001. Balancing Political and Ecological Values. *Environmental Politics* 10(1), 65–88.

Johnson, J. 2001. Arguing for Deliberation: Some Skeptical Considerations, in *Deliberative Democracy*, edited by J. Elster. Cambridge: Cambridge University Press, 161–84.

Johnson, J. 2001. "Argumentos en favor de la deliberación. Algunas consideraciones escépticas", en Jon Elster (ed.), *La democracia deliberativa*, Barcelona: Gedisa, 207–34.

Jordan III, W. 2003. *The Sunflower Forest: Ecological Restoration and the New Communion with Nature*. Berkeley: University of California Press.

Kallis, G. 2011. In Defence of Degrowth. *Ecological Economics* 70, 873–80.

Kant, I. 1986. *Kritike der reinen Vernunft*. Ditzingen: Reclam.

Karafyllis, N. 2003. "Das Wesen der Biofakte", in *Biofakte. Versuch über den Menschen zwischen Artefakf und Lebewesen*, edited by N. Karafyllis. Paderborn: Mentis, 11–26.

Kassman, K. 1997. *Envisioning Ecotopia. The U.S. Green Movement and the Politics of Radical Social Change*. Westport: Praeger.

Katz, E. 1985. Organism, Community, and the Substitution Problem. *Environmental Ethics* 7, 241–56.

Katz, E. 1992. The Big Lie: Human Restoration of Nature. *Research in Philosophy and Technology* 12, 231–41.

Katz, E. 1993. Artefacts and Functions: A Note on the Value of Nature. *Environmental Values* 2(3), 223–32.

Katz, E. 1997. *Nature as Subject. Human Obligation and Natural Community*. Lanham: Rowman and Littlefield Publishers.

Kidner, D. 2000. Fabricating Nature: A Critique of the Social Construction of Nature. *Environmental Ethics* 2(4), 339–58.

Kirkman, R. 2002. *Skeptical Environmentalism. The Limits of Philosophy and Science*. Bloomington and Indianapolis: Indiana University Press.

Kolakowski, L. 2006. *My Correct Views on Everything*. South Bend: St Augustine's Press.

Lafferty, W.M. and Meadowcroft, J. (eds) 1996. *Democracy and the Environment*. Cheltenham: Edward Elgar.

Latour, B. 2004. *Politics of Nature. How to Bring the Sciences into Democracy*. Cambridge: Harvard University Press.

Latta, A. 2007. Locating Democratic Politics in Ecological Citizenship. *Environmental Politics* 16(3), 377–93.

Lawn, P. 2010a. On the Ehrlich-Simon Bet: Both were Unskilled and Simon was Lucky. *Ecological Economics* 69, 2045–6.

Lawn, P. 2010b. Ecological Economics: The Impact of Unsustainable Growth, in *Routledge Handbook of Climate Change and Society*, edited by C. Lever-Tracy. London: Routledge, 94–120.

Leahy, T., Bowden, V. and Threadgold, S. 2010. Stumbling Towards Collapse: Coming to Terms with the Climate Crisis. *Environmental Politics* 19(6), 851–68.

Lease, G. 1995. Introduction: Nature under Fire, in *Reinventing Nature? Responses to Postmodern Deconstruction*, edited by M.E. Soulé and G. Lease. Washington, DC: Island Press, 3–16.

Lee, K. 1995. Beauty for Ever? *Environmental Values* 4, 213–25.

Lee, K. 2005. Is Nature Autonomous?, in *Recognizing the Autonomy of Nature. Theory and Practice*, edited by T. Heyd. New York: Columbia University Press, 54–74.

Leiss, W. 1994. *The Domination of Nature*. London: McGuill-Queen's University Press.

Lelé, S.M. 1991. *Sustainable Development: A Critical Review, World Development* 19(6), 607–21.

Leopold, A. 1987. *A Sand County Almanac. And Sketches Here and There*. Oxford: Oxford University Press.

Leroux, M. 2005. *Global Warming – Myth or Reality? The Erring Ways of Climatology*. Chichester: Springer and Praxis.

Lever-Tracy, C. and Pittock, B. 2010. Climate Change and Society: An Introduction, in *Routledge Handbook of Climate Change and Society*, edited by C. Lever-Tracy. London: Routledge, 1–10.

Levine, A. 1981. *Liberal Democracy: A Critique of its Theory*. New York: Columbia University Press.

Levy, Y. 2004. The End of Environmentalism (as we know it), in *Liberal Democracy and Environmentalism. The End of Environmentalism?*, edited by Y. Levy and M. Wissenburg. London: Routledge, 48–59.

Levy, Y. and Wissenburg, M. (eds) 2004a. *Liberal Democracy and Environmentalism. The End of Environmentalism?* London: Routledge.

Levy, Y. and Wissenburg, M. 2004b. Introduction, in *Liberal Democracy and Environmentalism. The End of Environmentalism?*, edited by Y. Levy and M. Wissenburg. London: Routledge, 1–9.

Levy, Y. and Wissenburg, M. 2004c. Conclusion, in *Liberal Democracy and Environmentalism. The End of Environmentalism?*, edited by Y. Levy and M. Wissenburg. London: Routledge, 193–6.

Lewis, M. 1992. *Green Delusions. An Environmentalist Critique of Radical Environmentalism*. Duke: Duke University Press.

Lidskog, R. 2000. Scientific Evidence or Lay People's Experience? On Risk and Trust with Regard to Modern Environmental Threats, in *Risk in the Modern Age. Social Theory, Science and Environmental Decision-Making*, edited by M. Cohen. London: Macmillan, 196–224.

Light, A. 2001. The Urban Blind Spot in Environmental Ethics. *Environmental Politics* 10, 7–35.

Light, A. 2005. Restoration, Autonomy, and Domination, in *Recognizing the Autonomy of Nature. Theory and Practice*, edited by T. Heyd. New York: Columbia University Press, 154–69.

Lomborg, B. 2001. *The Skeptical Environmentalist*. Cambridge: Cambridge University Press.

Luhmann, N. 1989. *Ecological Communication*. Cambridge: Polity.

Macedo, S. 1999. *Deliberative Politics. Essays on Democracy and Disagreement*. New York: Oxford University Press.

MacIntyre, A. 1988. *Whose Justice? Which Rationality?* Notre Dame: University of Notre Dame Press.

Macgregor, S. 2006. No Sustainability without Justice: A Feminist Critique of Environmental Citizenship, in *Environmental Citizenship*. Cambridge and London: The MIT Press, 101–26.

Macnaghten, P. and Urry, J. 1998. *Contested Natures*. London: Sage.

Macpherson, C.B. 1964. *The Political Theory of Possessive Individualism: Hobbes to Locke*. Oxford: Oxford Paperbacks.

Maddox, J. 1972. *The Doomsday Syndrome*. London: Macmillan.

Malthus, R. 1999. *An Essay on the Principle of Population*. New York: Oxford University Press.

Manin, B. 1987. On Legitimacy and Political Deliberation. *Political Theory* 15(3), 338–68.

Marshall, T.H. 1987. *Citizenship and Social Class*. London: Pluto Press.

Martell, L. 1994. *Ecology and Society. An Introduction*. Cambridge: Polity Press.

Martinsson, J. and Lundqvist, L. 2010. Ecological Citizenship: Coming Out 'Clean' without Turning 'Green'? *Environmental Politics* 19(4), 518–37.

Marx, K. 1993. *Capital: A Critique of Political Economy, vol. 3*. London: Penguin.

Marx, K. 2009. *Ökonomisch-philosophische Manuskripte*. Frankfurt: Suhrkamp.

Marx, K. and Engels, F. 2009. *Die Deutsche Ideologie*. Berlin: Akademie-Verlag.

Mathews, F. 1995a. Introduction. *Environmental Politics* 4(4), 1–12.

Mathews, F. 1995b. Community and the Ecological Self. *Environmental Politics* 4(4), 66–100.

McAnani, P. and Yoffee, N. (eds) 2010. *Questioning Collapse. Human Resilience, Ecological Vulnerability, and the Aftermath of Empire*. Cambridge: Cambridge University Press.

McKibben, B. 1990. *The End of Nature*. New York: Anchor Books.

McNeill, J.R. 2000. *Something New Under the Sun. An Environmental History of the Twentieth Century*. London: Penguin.

Meadows, D. and Meadows, D. 1972. *The Limits to Growth: A Report for the Club of Rome's Project on the Predicament of Mankind*. New York: Universe Books.

Melucci, A. 1996. *Challenging Codes. Collective Action in the Information Age*. Cambridge: Cambridge University Press.

Merchant, C. 1992. *Radical Ecology: The Search for a Livable World*. Routledge: New York.

Merchant, C. 1996. Reinventing Eden: Western Culture as a Recovery Narrative, in *Uncommon Ground. Rethinking the Human Place in Nature*, edited by W. Cronon. New York: W.W. Norton & Company, 132–59.

Meyer, J.M. 1999. Interpreting Nature and Politics in the History of Western Thought: The Environmentalist Challenge. *Environmental Politics* 8(2), 1–23.

Meyer, J.M. 2001. *Political Nature. Environmentalism and the Interpretation of Western Thought*. Cambridge and London: The MIT Press.

Meyer, S. 2006. *The End of the Wild*. Cambridge: MIT Press/Boston Review.

Midgley, M. 1995. *Beast and Man: The Roots of Human Nature*. London: Routledge.

Milbrath, L. 1984. *Environmentalists. Vanguard for a New Society*. New York: State University of New York Press.

Mills, M. 1996. Green Democracy: The Search for an Ethical Solution, in *Democracy and Green Political Thought*, edited by B. Doherty and M. de Geus. London: Routledge, 97–114.

Mills, M. 2001. The Duties of Being and Association, in *Sustaining Liberal Democracy. Ecological Challenges and Opportunities*, edited by J. Barry and M. Wissenburg. Houndmills: Palgrave, 163–78.

Mills, M. and King, F. 2004. The End of Deep Ecology? – Not Quite, in *Liberal Democracy and Environmentalism. The End of Environmentalism?*, edited by Y. Levy and M. Wissenburg. London: Routledge, 75–86.

Minteer, B. 2002. Deweyan Democracy and Environmental Ethics, in *Democracy and the Claims of Nature. Critical Perspectives for a New Century*, edited by B. Minteer and B. Pepperman Taylor. Lanham: Rowman and Littlefield Publishers, 33–48.

Minteer, B. and Pepperman Taylor, B. (ed.) 2002. *Democracy and the Claims of Nature. Critical Perspectives for a New Century*. Lanham: Rowman and Littlefield Publishers.

Moffat, I. 2007. Environmental Space, Material Flow Analysis and Ecological Footprinting, in *Handbook of Sustainable Development*, edited by G. Atkinson, S. Dietz and E. Neumayer. Cheltenham and Northampton: Edward Elgar, 319–44.

Mol, A. 2008. *Environmental Reform in the Information Age*. Cambridge University Press: Cambridge.

Mol, A., Sonnenfeld, D. and Spaargaren, G. (eds) 2009. *The Ecological Modernisation Reader*. London: Routledge.

Moriarty, P. 2007. Nature Naturalized: A Darwinian Defense of the Nature/Culture Distinction, *Environmental Ethics* 29, 227–46.

Moscovici, S. 1975. *Sociedad contra Natura*, México DF: Siglo XXI. (English edition: *Society against Nature: The Emergence of Human Societies*. Sussex: Harvester Press, 1976.)

Murtaza, N. 2011. Pursuing Self-interest or Self-actualization? From Capitalism to a Steady-state, Wisdom Economy. *Ecological Economics* 70, 577–84.

Musschenga, B. 1994. Liberal Neutrality and the Justification of Environmental Conservation, in *Ecology, Technology and Culture*, edited by W. Zeers and J. Boersema. Cambridge: The White Horse Press, 164–74.

Naess, A. 1989. *Ecology, Community and Lifestyle*. Cambridge: Cambridge University Press.

Neumayer, E. 2010. *Weak versus Strong Sustainability. Exploring the Limits of Two Opposing Paradigms*, 3rd edition. Cheltenham: Edward Elgar.

Neuteleers, S. 2010. Institutions versus Lifestyle: Do Citizens Have Environmental Duties in their Private Spheres? *Environmental Politics* 19(4), 501–17.

Newton, T. 2007. *Nature and Sociology*. London: Routledge.

Niemeyer, S. 2004. Deliberation in the Wilderness: Displacing Symbolic Politics. *Environmental Politics* 13(2), 347–72.

Nordhaus, T. and Shellenberg, M. 2007. *Break Through. From the Death of Environmentalism to the Politics of Possibility*. Boston: Houghton Mifflin.

Norton, B. 1991. *Toward Unity Among Environmentalists*. Oxford: Oxford University Press.

Norton, B. 2002. Democracy and Environmentalism: Foundations and Justifications in Environmental Policy, in *Democracy and the Claims of Nature. Critical Perspectives for a New Century*, edited by B. Minteer and B. Pepperman Taylor. Lanham: Rowman and Littlefield Publishers, 11–32.

Norton, B. 2005. *Sustainability. A Philosophy of Adaptative Ecosystem Management*. Chicago: The University of Chicago Press.

Nozick, R. 1984. *Philosophical Explanations*. Oxford: Oxford University Press.

Nozick, R. 2008. *Anarchy, State, and Utopia*. Malden: Blackwell.

Oelschlager, M. 1991. *The Idea of Wilderness. From Prehistory to the Age of Ecology*. New Haven: Yale University Press.

O'Neill, J. 1992. The Varieties of Intrinsic Value. *The Monist* 75(2), 119–37.

O'Neill, J. 1993. *Ecology, Policy and Politics*. London: Routledge.

O'Neill, J. 2009. Sustainability, Welfare and Value over Time, in *Governing Sustainability*, edited by W.N. Adler and A. Jordan. Cambridge: Cambridge University Press, 283–304.

O'Neill, J., Turner, K. and Bateman, I. (eds) 2001. *Environmental Ethics and Philosophy*. Cheltenham: Edward Elgar.

Ophuls, W. 1977. *Ecology and the Politics of Scarcity*. San Francisco: W.H. Freeman and Company.

Opschoor, J.B. and Weterings, R. 1994. Environmental Utilisation Space: An Introduction. *Milieu* 9(4), 198–205.

O'Riordan, T. 1976. *Environmentalism*. London: Pion Limited.

O'Riordan, T. 2009. Reflection on the Pathways to Sustainability, in *Governing Sustainability*, edited by W.N. Adler and A. Jordan. Cambridge: Cambridge University Press, 307–28.

Ott, K. 2009. On Substantiating the Conception of Strong Sustainability, in *Sustainability, Natural Capital and Nature Conservation*, edited by R. Döring. Marburg: Metropolis-Verlag, 49–72.

Ott, K. and Döring, R. 2004. *Theorie und Praxis starker Nachhaltigkeit*. Marburg: Metropolis-Verlag.

Pascal, B. 1995. *Pensées*. London: Penguin.

Passmore, J. 1974. *Man's Responsibility for Nature: Ecological Problems and Western Traditions*. New York: Scribner.

Passmore, J. 2000. *The Perfectibility of Man*, 3rd edition. Indianapolis: Liberty Fund.

Patt, A. et al. 2010. What Can Social Science Tell Us about Meeting the Challenge of Climate Change? Five Insights from Five Years that Might Make a Difference, in *Making Climate Change Work for Us: European Perspectives on Adaptation and Mitigation Strategies*, edited by M. Hulme and H. Neufeldt. Cambridge: Cambridge University Press, 369–88.

Patterson, M. 2009. Global Governance for Sustainable Capitalism? The Political Economy of Global Environment, in *Governing Sustainability*, edited by W.N. Adler and A. Jordan. Cambridge: Cambridge University Press, 99–122.

Pearce, F. 2007. *With Speed and Violence: Why Scientists Fear Tipping Points in Climate Change*. Uckfield: Beacon Press.

Pennington, Mark. 2003. Hayekian Political Economy and the Limits of Deliberative Democracy. *Political Studies* 51, 722–39.

Pepper, D. 1984. *The Roots of Modern Environmentalism*. London: Routledge.

Pepper, D. 1993a. *Eco-socialism. From Deep Ecology to Social Justice*. London: Routledge.

Pepper, D. 1993b. Anthropocentrism, Humanism and Eco-Socialism: A Blueprint for the Survival of Ecological Politics. *Environmental Politics* 2(3), 428–52.

Pepper, D. 1996. *Modern Environmentalism. An Introduction*. London: Routledge.

Pepperman Taylor, B. and Minteer, B. (eds) 2002. *Democracy and the Claims of Nature. Critical Perspectives for a New Century*. Lanham: Rowman and Littlefield Publishers.

Perler, D. and Wild, M. 2005. Der Geist der Tiere – eine Einführung, in *Der Geist der Tiere. Philosophische Texte zu einer aktuellen Diskussion*, edited by D. Perler and M. Wild. Frankfurt: Suhrkamp, 10–74.

Peterson, A. 1999. Environmental Ethics and the Social Construction of Nature. *Environmental Ethics* 21, 339–57.

Pinker, S. 2002. *The Blank Slate. The Modern Denial of Human Nature*. London: Penguin.

Plumwood, V. 1993. *Feminism and the Mastery of Nature*. London: Routledge.

Plumwood, V. 2002. *Environmental Culture. The Ecological Crisis of Reason*. London: Routledge.

Plumwood, V. 2005. Toward a Progressive Naturalism, in *Recognizing the Autonomy of Nature. Theory and Practice*, edited by T. Heyd. New York: Columbia University Press, 25–53.

Ponthiere, G. 2009. The Ecological Footprint: An Exhibit at an Intergenerational Trial? *Environment, Development, Sustainability* 11, 677–94.

Ponting, C. 2007. *A New Green History of the World. The Environment and the Collapse of Great Civilisations*. London: Vintage.

Porritt, J. 1984. *Seeing Green. The Politics of Ecology Explained*. London: Basil Blackwell.

Porritt, J. 2005. *Capitalism as if the World Matters*. London: Earthscan.

Price, J. 1998. Looking for Nature at the Mall: A Field Guide to the Nature Company, in *Uncommon Ground. Rethinking the Human Place in Nature*, edited by W. Cronon. New York: W.W. Norton & Company, 186–203.

Princen, T. 2010. *Treading Softly: Paths to Ecological Order*. Cambridge: MIT Press.

Radcliffe, J. 2000. *Green Politics. Dictatorship or Democracy?* London: Macmillan.

Radkau, J. 2000. *Natur und Macht. Eine Weltgeschichte der Umwelt*. Munich: C.H. Beck.

Rawls, J. 1993. *Political Liberalism*. New York and Chichester: Columbia University Press.

Raz, J. 1988. *The Morality of Freedom*. Oxford: Oxford University Press.

Redclift, M. 1987. *Sustainable Development. Exploring the Contradictions*. New York: Methuen.

Redclift, M. 1999a. Pathways to Sustainability: Issues, Policies and Theories, in *Planning Sustainability*, edited by M. Kenny and J. Meadowcroft. London: Routledge, 66–77.

Redclift, M. 1999b. Sustainability and Sociology: Northern Preoccupations, in *Sustainability and the Social Sciences. A Cross-disciplinary Approach to Integrating Environmental Considerations into Theoretical Reorientation*, edited by J. Becker et al. London: Zed Books, 59–73.

Redclift, M. 2006. *Frontiers. Histories of Civil Society and Nature*. Cambridge: MIT Press.

Regan, T. 2004. *The Case for Animal Rights*. Berkeley: University of California Press.

Renn, O. 2009. Precaution and the Governance of Risk, in *Governing Sustainability*, edited by W.N. Adler and A. Jordan. Cambridge: Cambridge University Press, 226–58.

Retallack, S. and Lawrence, T. 2007. *Positive Energy: Harnessing People Power to Prevent Climate Change*. London: Institute of Public Policy Research.

Reyner, T. and Okereke, C. 2007. The Politics of Climate Change, in *The Politics of the Environment: A Survey*, edited by C. Okereke. London: Routledge, 116–35.

Ridley, M. 2004. *Nature via Nurture. Genes, Experience and What Makes Us Human*. London: Harper Perennial.

Ridley, M. 2010. *The Rational Optimist*. London: Fourth State.

Rogers, T. 2009. Nature of the Third Kind: Toward an Explicitly Relational Constructionism. *Environmental Ethics* 31, 393–412.

Rosset, C. 1974. *La anti-naturaleza*. Madrid: Taurus.

Roughley, N. 2005. Was heißt 'menschliche Natur'? Begriffliche Differenzierungen und normative Ansatzpunkte, in *Die Menschliche Natur. Welchen und wiviel Wert hat sie?*, edited by K. Bayertz. Paderborn: Mentis, 133–56.

Russell, E. 2011. *Evolutionary History: Uniting History and Biology to Understand Life on Earth*. Cambridge: Cambridge University Press.

Ruta, G. and Hamilton, K. 2007. The Capital Approach to Sustainability, in *Handbook of Sustainable Development*, edited by G. Atkinson, S. Dietz and E. Neumayer. Cheltenham and Northampton: Edward Elgar, 45–62.

Sagoff, M. 1988. *The Economy of the Earth. Philosophy, Law and the Environment*. Cambridge: Cambridge University Press.

Sagoff, M. 2004. *Price, Principle, and the Environment*. Cambridge: Cambridge University Press.

Sale, K. 1985. *Dwellers in the Land. The Bioregional Vision*. San Francisco: Sierra Club Books.

Sale, K. 2006. *After Eden: The Evolution of Human Domination*. Durham: Duke University Press.

Sánchez Ferlosio, R. 1990. Comentarios del traductor, in *Victor de L'Aveyron* by Jean Itard. Madrid: Alianza.

Sánchez Ferlosio, R. 2009. *'Guapo'y sus isótopos*. Barcelona: Destino.

Sanders, L. 1997. Against Deliberation. *Political Theory* 25(3), 347–76.

Sartori, G. 2002. *Elementos de teoría política*. Madrid: Alianza.

Saurin, J. 2001. Global Environmental Crisis as the 'Disaster Triumphant': The Private Capture of Public Goods. *Environmental Politics* 10(4), 63–84.

Saward, M. 1996. Must Democrats be Environmentalists?, in *Democracy and Green Political Thought. Sustainability, Rights and Citizenship*, edited by B. Doherty and M. de Geus. London: Routledge, 79–96.

Saward, M. 1998. *The Terms of Democracy*. Cambridge: Polity.

Saward, M. (ed.) 2000. *Democratic Innovation. Deliberation, Representation and Association*. London: Routledge.

Schmidt, A. 1971. *The Concept of Nature in Marx*. Londres: NLB.

Schmidt, G. 2005. *Positive Ecology. Sustainability and the 'Good Life'*. Aldershot: Ashgate.

Schor, J. 2010. *Plentitude: The New Economics of Wealth*. New York: Penguin.

Seyfang, G. 2006. Shopping for Sustainability: Can Sustainable Consumption Promote Ecological Citizenship?, in *Citizenship, Environment, Economy*, edited by A. Dobson and A. Valencia. London: Routledge, 137–53.

Shearman, D. and Smith, J.W. 2007. *The Climate Change Challenge and the Failure of Democracy*. Westsport: Praeger.

Sheldrake, R. 1990. *The Rebirth of Nature. The Greening of Science and God*. London: Rider.

Siep, L. 2004. *Konkrete Ethik. Grundlagen der Natur- und Kulturethik*. Frankfurt: Suhrkamp.

Sikor, T. and Norggard, R. 1999. Principles for Sustainability: Protection, Investment, Co-operation and Innovation, in *Sustainability in Question. The Search for a Conceptual Framework*, edited by J. Köhn et al. Cheltenham: Edward Elgar, 49–65.

Simon, J. 1981. *The Ultimate Resource*. Princeton: Princeton University Press.

Simon, J. (ed.) 1995. *The State of Humanity*. Cambridge: Blackwell.

Simon, J. and Kahn, H. 1984. *The Resourceful Earth*. Oxford: Wiley-Blackwell.

Singer, P. 1976. *Animal Liberation. Towards an End to Man's Inhumanity to Animals*. London: Jonathan Cape.

Sismondo, S. 1999. Some Social Constructions. *Social Studies of Science* 23, 515–53.

Sloterdijk, P. 1999. *Regeln für den Menschenpark. Ein Antwortschreiben zu Heideggers Brief über den Humanismus*. Frankfurt: Suhrkamp.

Smail, D.L. 2008. *Deep History and the Brain*. Berkeley and Los Angeles: University of California Press.

Smith, G. 2000. Toward Deliberative Institutions, in *Democratic Innovation. Deliberation, Representation and Association*, edited by M. Saward. London: Routledge, 29–39.

Smith, G. 2001. Taking Deliberation Seriously: Institutional Design and Green Politics. *Environmental Politics* 10(3), 72–93.

Smith, G. 2003. *Deliberative Democracy and the Environment*. London: Routledge.

Smith, M.J. 1998. *Ecologism. Towards Ecological Citizenship*. Buckingham: Open University Press.

Smith, M. 1999. To Speak of Trees: Social Constructivism, Environmental Values, and the Future of Deep Ecology. *Environmental Ethics* 21, 359–76.

Snow, D. 2004. Framing Processes, Ideology, and Discursive Fields, in *The Blackwell Companion to Social Movements*, edited by D. Snow, S. Soule and H. Kriesi. Oxford: Blackwell, 3–16.

Soper, K. 1995. *What is Nature?* Oxford: Blackwell.

Spen Stokes, P. 2009. *Money & Soul. The Psychology of Money and the Transformation of Capitalism*. Devon: Green Books.

Steenbergen, B. 1994. Towards a Global Ecological Citizenship, in *The Condition of Citizenship*, edited by B. Steenbergen. London: Sage, 41–152.

Steffen, W., Grinevald, J., Crutzen, P. and McNeill, J. 2011. The Anthropocene: Conceptual and Historical Perspectives. *Philosophical Transactions of the Royal Society* 369, 842–67.

Stephens, P. 2000. Nature, Purity, Ontology. *Environmental Values* 9, 267–94.

Stephens, P. 2001. Green Liberalism: Nature, Agency and the Good. *Environmental Politics* 10(3), 1–22.

Stone, C.D. 2010. *Should Trees Have Standing? Toward Legal Rights for Natural Objects*. Oxford: Oxford University Press.

Stoner, J. and Wankel, C. 2008. Introduction: Exploring New Frameworks, Practices, and Initiatives for a Sustainable World, in *Innovative Approaches to Global Sustainability*, edited by C. Wankel and J. Stoner. New York: Palgrave Macmillan, 3–10.

Stuart Mill, J. 1998. *Three Essays on Religion*. Amherst: Prometheus Books.

Stuart Mill, J. 2001. *Utilitarianism*. Indianapolis: Hackett.

Sunstein, C. 2006. *Infotopia. How Many Minds Produce Knowledge*. New York: Oxford University Press.

Sunstein, C. and Hastie, R. 2008. *Four Failures of Deliberating Groups*. John M. Olin Law & Economics Working Paper 401.

Surowiecki, J. 2004. *The Wisdom of Crowds. Why the Many are Smarter than the Few*. New York: Anchor Books.

Sutton, P.W. 2004. *Nature, Environment and Society*. Houndmills: Palgrave Macmillan.

Talisse, R. 2004. Does Public Ignorance Defeat Deliberative Democracy? *Critical Review* 16, 455–63.

Talshir, G. 2004. The Role of Environmentalism: From 'The Silent Spring' to 'The Silent Revolution', in *Liberal Democracy and Environmentalism. The End of Environmentalism?*, edited by Y. Levy and M. Wissenburg. London: Routledge, 10–31.

Tapscott, D. and Williams, A. 2008. *Wikinomics. How Mass Collaboration Changes Everything*. London: Atlantic Books.

Tapscott, D. and Williams, A. 2010. *Macrowikinomics: Rebooting Business and the World*. London: Atlantic Books.

Thaler, R. and Sunstein, C. 2009. *Nudge. Improving Decisions about Health, Wealth and Happiness*. London: Penguin.

The Economist. 2007. Jolly Green Heretic. 8–14 September, *Technology Quarterly* 27–8.

The Hartwell Group. 2010. *The Hartwell Paper. A New Direction for Climate Policy After the Crash of 2009*. University of Oxford/London School of Economics.

Thomas, K. 1984. *Man and the Natural World*. London: Penguin.

Tomlinson, B. 2010. *Greening through IT. Information Technology for Environmental Sustainability*. Cambridge and London: The MIT Press.

Torgerson, D. 1995. The Uncertain Quest for Sustainability: Public Discourse and the Politics of Environmentalism, in *Greening Environmental Policy: The Politics of a Sustainable Future*, edited by F. Fischer and M. Black. London: Paul Chapman, 3–20.

Torgerson, D. 1999. *The Promise of Green Politics*. Durham: Duke University Press.

Trachtenberg, Z. 2010. Complex Green Citizenship and the Necessity of Judgement. *Environmental Politics* 19(3), 339–55.

Trainer, T. 2007. *Renewable Energy Cannot Sustain a Consumer Society*. Dordrecht: Springer.

Turner, B. 1994. Postmodern Culture/Modern Citizens, in *The Condition of Citizenship*, edited by B. van Steenbergen. London: Sage, 153–68.

Urry, J. 2008. Climate Change, Travel and Complex Futures. *British Journal of Sociology* 59(2), 261–79.

Valencia, A. 2005. Globalisation, Cosmopolitanism and Ecological Citizenship, *Environmental Politics* 14(2), 163–78.

Valencia, A., Arias-Maldonado, M. and Vázquez, R. 2010. *Ciudadanía y conciencia medioambiental en España*. Madrid: Centro de Investigaciones Sociológicas.

Vallespín, F. 2000. *El futuro de la política*. Madrid: Taurus.

Van den Bergh, J. 2007. Sustainable Development in Ecological Economics, in *Handbook of Sustainable Development*, edited by G. Atkinson, S. Dietz and E. Neumayer. Cheltenham and Northampton: Edward Elgar, 63–77.

Van den Bergh, J. 2011. Environment versus Growth – A Criticism of 'Degrowth' and a Plea for 'A-Growth'. *Ecological Economics* 70, 881–90.

Vanderheiden, S. 2011. Rethinking Environmentalism: Beyond Doom and Gloom. *Global Environmental Politics* 11(1), 108–13.

Victor, P.A. 1991. Indicators of Sustainable Development: Some Lessons from Capital Theory. *Ecological Economics* 4, 191–213.

Victor, P.A. 2008. *Managing without Growth. Slower by Design, Not Disaster*. Cheltenham: Edward Elgar.

Vincent, A. 1998. Liberalism and the Environment. *Environmental Values* 7, 443–59.

Vogel, S. 2002. Environmental Philosophy After the End of Nature. *Environmental Ethics* 24, 23–39.

Von Mises, L. 2007. *Human Action. A Treatise on Economics*. Indianapolis: Liberty Fund.

Von Uexküll, J. 2010. *Bausteine Zu Einer Biologischen Weltanschauung*. Wiltshire: Bertrams.

Von Uexküll, J. and Kriszat, G. 1970. *Streifzüge durch die Umwelten von Tieren und Menschen*. Bedeutungslehre. Frankfurt: Fischer Verlag.

Wackernagel, M. and Rees, W.E. 1996. *Our Ecological Footprint: Reducing Human Impact on the Earth.* Gabriola Island and Philadelphia: New Society Publishers.

Wapner, P. 2010. *Living Through the End of Nature: The Future of American Environmentalism.* Cambridge: MIT Press.

Weale, A. 2009. Governance, Government and the Pursuit of Sustainability, in *Governing Sustainability*, edited by W.N. Adler and A. Jordan. Cambridge: Cambridge University Press, 55–75.

Welzer, H. 2008. *Klimakriege. Wofür im 21. Jahrhundert getötet wird.* Frankfurt am Main: Fischer Verlag.

Westra, L. 1993. The Ethics of Environmental Holism and the Democratic State: Are They in Conflict? *Environmental Values* 2, 125–36.

Williams, R. 1980. *Problems in Materialism and Culture. Selected Essays.* London: Verso.

Wilson, E.O. 1975. *Sociobiology: The New Synthesis.* Cambridge: Harvard University Press.

Wilson, E.O. 2003. The Encyclopedia of Life. *Trends in Ecology and Evolution* 18(2), 77–80.

Wissenburg, M. 1998. *Green Liberalism. The Free and the Green Society.* London: UCL Press.

Wissenburg, M. 2004. Little Green Lies: On the Redundancy of 'Environment', in *Liberal Democracy and Environmentalism. The End of Environmentalism?*, edited by Y. Levy and M. Wissenburg. London: Routledge, 60–71.

World Commission on Environment and Development 1987. *Our Common Future.* Oxford: Oxford University Press.

Young, I.M. 1996. Communication and the Other: Beyond Deliberative Democracy, in *Democracy and Difference. Contesting the Boundaries of the Political*, edited by S. Benhabib. Princeton: Princeton University Press, 120–35.

Young, J. 1990. *Post-environmentalism.* London: Belhaven Press.

Index